Environmental Sensing

Environmental Sensing

Editor

Atul Singh

scitus
academics

Environmental Sensing

Edited by **Atul Singh**

ISBN: 978-1-68117-138-8

Library of Congress Control Number: 2015936558

SCITUS Academics LLC,
www.scitusacademics.com
Box No. 4766, 616 Corporate Way,
Suite 2, Valley Cottage,
NY 10989

Printed in United States of America on Acid Free Paper ∞

Contents

Preface

Remote sensing is neither a difficult or exotic technology. At its most fundamental level, remote sensing is a form of basic field data collection, an activity common to all physical, natural, and social science and motivated by the same overarching goals: to systematically observe phenomena in order to record useful measurements of the variables that define their characteristic properties. Remote sensing simply acquires those measurements in a very different manner. Anyone with a background in subject areas ranging from geography, Earth science, planning, or resource management, to perhaps less obvious fields such as public policy studies, environmental health, and international development can apply remote sensing methods successfully and produce useful information that can address a myriad of problems and issues. To approach this wider audience, I have elected to title this book using the term "environmental sensing," where the verb sensing means "to become aware of"; aware of the technology and of the environmental patterns this technology can illuminate.

Editor

Chapter 1

Empirical Demonstration of Environmental Sensing in Catalytic RNA: Evolution of Interpretive Behavior at the Origins of Life

Niles Lehman[1], Tess Bernhard[2], Brian C Larson[1], Andrew JN Robinson[3], and Christopher CB Southgate[3]

[1]Department of Chemistry, Portland State University, Portland, OR, USA

[2]Department of Ecology and Evolutionary Biology, Princeton University, Princeton, NJ, USA

[3]College of Humanities, University of Exeter, Exeter, UK

ABSTRACT

Background

The origins of life on the Earth required chemical entities to interact with their environments in ways that could respond to natural selection.

The concept of interpretation, where biotic entities use signs in their environment as proxy for the existence of other items of selective value in their environment, has been proposed on theoretical grounds to be relevant to the origins and early evolution of life. However this concept has not been demonstrated empirically.

Results

Here, we present data that certain catalytic RNA sequences have properties that would enable interpretation of divalent cation levels in their environment. By assaying the responsiveness of two variants of the Tetrahymena ribozyme to the Ca^{2+} ion as a sign for the more catalytically useful Mg^{2+}ion, we show an empirical proof-of-principle that interpretation can be an evolvable trait in RNA, often suggested as a model system for early life. In particular we demonstrate that in vitro, the wild-type version of the Tetrahymena ribozyme is not interpretive, in that it cannot use Ca^{2+} as a sign for Mg^{2+}. Yet a variant of this sequence containing five mutations that alter its ability to utilize the Ca^{2+}ion engenders a strong interpretive characteristic in this RNA.

Conclusions

We have shown that RNA molecules in a test tube can meet the minimum criteria for the evolution of interpretive behaviour in regards to their responses to divalent metal ion concentrations in their environment. Interpretation in RNA molecules provides a property entirely dependent on natural physico-chemical interactions, but capable of shaping the evolutionary trajectory of macromolecules, especially in the earliest stages of life's history.

BACKGROUND

The origins of life on the Earth can be described as a transition from an environment containing chemicals dominated by mass-action kinetics to one containing informational macromolecules dominated by selective forces and evolution [1],[2]. The latter phenomena involve complex molecular 'behaviour' whereby populations of molecules compete for reproductive success. The interaction of organisms with an ever-changing environment has been a key evolutionary factor since the earliest entities emerged that were subject to natural selection [3]. It is thus of great importance to the understanding of the origin of life to understand the nascent stages of this interaction.

For a genotype (not necessarily originally a complete genome) to interact with its environment, it must first sense, and then respond to, environmental states. Primordial genotypes would not have been adept at these actions, as they would have not had the benefit of extensive evolutionary tuning. We contend that the most characteristic form of this sensing in living organisms (as opposed to inert material responding mechanically) is an interpretive process involving a response to a sign. At the molecular level, signs are indicators of a more general environmental condition; in particular here we will focus on metal ions as indicative of the presence or absence of other ions in the environment as a consequence of causally related solubilities in an aqueous milieu.

The idea that interpretation, representation, or signification is a fundamental characteristic of living things is the basis of the field of theoretical biology known as biosemiotics (semiotics being the philosophical study of signs) [4]. Biosemiotics is a controversial area of study because it has tended to lack empirically testable propositions. Our focus here is on providing just such opportunities for testing. Originally genotypes would be poor at correctly interpreting the state of the environment, but this trait would be subject to natural selection. To help explore how this evolution could happen, what was first needed was a conceptually robust definition of interpretation that cannot be reduced to a fully adequate re-description in terms only of mechanistic causes and effects. Though the interactions are entirely mechanistic when considered at the molecular level of description, a fully adequate explanation of why such properties would be selected and maintained by selection requires, in addition, the semiotic level of description. Two of the present authors have proposed such a definition [5].

In an interpretation, the trigger for the change of state (the 'sign') is related to, but not obligatorily implied by, some aspect of the environment (the 'object') that would make such a change in state beneficial. For example, an amoeba will crawl up a chemical gradient of attractant molecules towards the latter's source. The chemotactic response is adaptive because it increases the chance of the amoeba obtaining nutrition (by ingesting the bacterium that is the source of the attractant). The response is interpretive because it is not the chemical gradient that is of benefit to the amoeba, but the bacterium of which the attractant molecules are a sign. The amoeba's target (the bacterium) is physically distinct from the sign (the chemical attractant). It is this separation between sign and object that distinguishes an interpretation from a direct adaptive response; i.e., interpretations arefallible. The amoeba may swim up a chemical gradient that has not, in a particular case, been produced by a potential source of nutrition. In general, interpretive behaviour may be adaptive in spite of its fallibility because on balance the benefits derived from obtaining information about the object without having to respond directly to the object outweigh the costs of misinterpretations. The amoeba benefits from responding the attractant gradient because it can thereby detect the presence of a bacterium at a distance.

The simplest conceivable scenario in which an adaptive interpretive response could occur would involve a two-state entity in a two-state environment [5]. Suppose that the environment can be 'favourable' (F) or 'unfavourable' (U) and the entity has two possible states, A and B. Suppose further that in environment F it is advantageous for the entity to be in state A, and in environment U it is advantageous for the entity to be in state B. The overall 'fitness' of the entity in this varying environment may be expressed, by analogy with game theory, in terms of a 2×2 payoff matrix (Figure 1). There are four possible outcomes in the matrix: O1 (environment F, entity state A), O2(environment U, entity state A), O3 (environment F, entity state B), and O4 (environment U, entity state B). The total 'payoff' (overall fitness) for the entity is the sum of O1 to O4, weighted according to the relative probabilities of each of these outcomes. In the non-interpretive configuration of the entity, its state (A or B) varies independently of the state of the environment. An interpretive variant of the entity might be capable of, say, responding to some sign that indicates (fallibly) that the state of the environment is F, the response being a change from state B to state A. This variant may have a selective advantage over the wildtype because it will increase the time that it spends in the advantageous combination of environment F and state A. Such interpretive responsiveness will only be adaptive, however, if it is not outweighed by the costs of a misinterpretation. This cost will be a function of the degree of disadvantage entailed by the entity being in state A in environment U, and the probability of being so placed in such a relation to the environment by the fallible (i.e., less than perfect) correlation between the presence of the sign and environmental state F.

		environmental condition	
		Favorable 25mM Mg^{2+}	Unfavourable 0.5 mM Mg^{2+}
entity state	State A 10mM Ca^{2+}	O1	O2
	State B no Ca^{2+}	O3	O4

Figure 1: Example payoff matrix for interpretive behaviour. Payoff values O1–O4 are evaluated for each pair-wise combination of environmental conditions and genotype traits as discussed in the text. The ion concentrations refer to those used in the assays of the Tetrahymenaribozyme, as described in Figures 2, 3 and 4.

For responsiveness to the sign to have the potential to count as an interpretation it is thus necessary that there be both a potential advantage to the response in the appropriate environment, and a potential disadvantage if the response occurs in the inappropriate environment (a misinterpretation). This may be expressed in terms of three inequalities that together provide the minimum criteria for interpretive behaviour in a 2-environment, 2-states-of-an-entity system. The relevant inequalities are:O1 > O2 (in state A the entity is better off in the favourable rather that the unfavourable environment); O1 > O3 (in the favourable environment the entity is better off in state A than state B);O4 > O2 (in the unfavourable environment the entity is better off in state B than state A, hence thecost of a misinterpretation) [6]. In addition, by definition, the highest of the four possible outcomes must correspond to the nominally favourable environment, otherwise that environment cannot be said to be the favourable one.

With this theoretical framework in hand, it then becomes imperative to demonstrate these phenomena in an empirical system. An obvious choice to manifest interpretation experimentally is ribonucleic acid (RNA), because RNA molecules simultaneously possess evolvability and catalytic function, and the RNA World concept is currently generally accepted as a plausible stage of early life on the Earth [1],[7]. Specifically, we wished to test the hypothesis that a single biomolecule not part of a more complex cellular system may exhibit properties consistent with the minimum criteria for interpretive responsiveness. In this paper we report investigations of the properties of the group I self-splicing ribozyme from the ciliated protozoan Tetrahymena. As sign and object we use the availability of the divalent cations (M^{2+}) which are not only prebiotically relevant but known to affect the catalytic behavior of ribozymes [8]-[10], and the response was a change in the catalytic rate of the ribozyme. We show that the cation dependency of the catalytic properties of a variant of this ribozyme could, under appropriate conditions in a hypothetical two-state environment, provide the basis on which the inequalities required as minimal conditions for interpretive responsiveness could be satisfied. We also demonstrate that such interpretive responsiveness can be an evolvable property in variants of this ribozyme, particularly in an origin-of-life context.

RESULTS

Pick-up-the-tail Assays

We chose two variants of the L–21 Tetrahymena ribozyme for study (Figure 2). The contemporary genotype (here, CG), an example of a present-day 'wildtype', is the L–21 derivative of the naturally occurring sequence obtained from T. thermophila and used in one of the first published examples of a catalytic RNA [11]. This ribozyme requires divalent cations for self-splicing or trans-splicing activityin vitro, and can utilize Mg^{2+} and/or Mn^{2+} but no other [8]. The 'primitive variant' (here, PV) is a five-error mutant of the wild-type ribozyme that was discovered through in vitro selection for activity in 10 mM $CaCl_2$ (clone 61 in ref. [12]). This variant has a catalytic efficiency in pure $CaCl_2$ that is five orders of magnitude less than that of the wild-type ribozyme (i.e., the CG) in $MgCl_2$, but its slight activity in Ca^{2+} is nonetheless notable because the ionic radius of Ca^{2+} is significantly larger than that of Mg^{2+}, and the five mutations are required to open up the active site enough to confer detectable catalytic activity [10]. We tracked the ability of both the CG and PV versions of this ribozyme to catalyze phospho-ester transfer in four cation-specific buffer conditions. First we measured activity as a function of [$MgCl_2$] and noted that there was a point of maximum catalytic separation between the two variants at 0.5 mM $MgCl_2$ (Figure 3). From this we chose 'high' and 'low' concentrations of Mg to use in subsequent assays; these values were 25 mM and 0.5 mM respectively. These values have been shown before as being above and below, respectively, the midpoint of the Tetrahymena activity assay [8]. We chose 10 mM $CaCl_2$ as the concentration of this alternative divalent (here, the sign) because this quantity is known to enhance folding of the wild-type Tetrahymena ribozyme in the presence or absence of $MgCl_2$[8] and is also the concentration at which the PV ribozyme was selected for activity in the absence of $MgCl_2$[12]. These ionic concentrations provide the clearest distinction between the properties of the two ribozymes and thus should optimize our chances of detecting interpretive behavior as an empirical proof-of-principle.

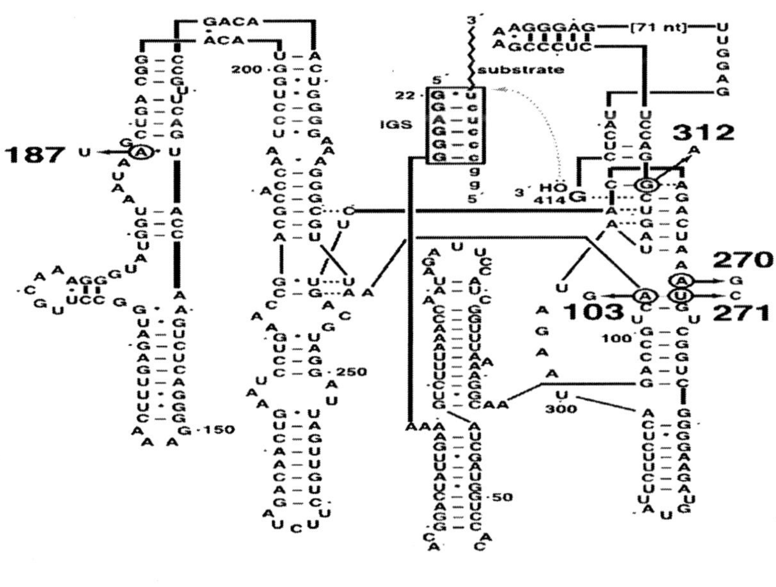

Figure 2: TheTetrahymenaribozymes, with secondary structures modified from [[10]]. The L–21 ('length minus 21' nucleotides from the 5' end of the

in vivo intron) sequence is shown for the wild-type ribozyme, with the five mutations in the $CaCl_2$-competent variant (PV) [12] indicated by the circles: A103G, A187U, A270G, U271C, and G312A. Numbering scheme follows the original [8],[11],[12], with dots provided every ten nucleotides. The interaction between the internal guide sequence (IGS) and six nucleotides of the substrate is shown by the grey box; in the 'pick-up-the-tail' assay the hydroxyl of the 3′ nucleotide (G414, shown) attacks at the splice site in the substrate (lower-case letters) and transfers the 3′ portion of the substrate (the last 17 nucleotides of S-1t in this case = AAAUAAAUAAAUAAAUA) to the 3′ end of the ribozyme, thereby lengthening it making it detectable by gel electrophoretic analysis. In this drawing, 71 nucleotide near the 3′ end of the ribozyme were omitted for clarity. Bars between nucleotide pairs denote canonical Watson-Crick pairs, while dots denote non-canonical base-paring interactions. The active site of this ribozyme is the environment around the G414 and the stack of base-triples that is above and below it [13].

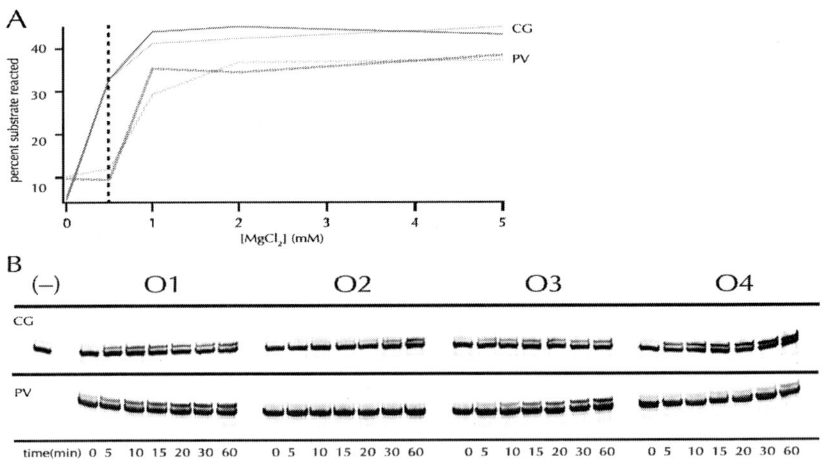

Figure 3: Assays of the ribozymes in various divalent ion concentrations. (A) Preliminary pick-up-the-tail assays were performed with 2.5-fold substrate excess as described in the text for activity after 60 minutes at 37°C. (Note: the original selection of the PV ribozyme [12] was performed at 3.25 h, at which time the PV activity surpasses that of the CG activity). The fraction of the initial amount of ribozyme RNA that reacted under conditions of 2.5-fold excess substrate was measured for each variant in the absence (green lines) or presence (red lines) of 10 mM $CaCl_2$ under concentrations of $MgCl_2$ ranging from 0.5–25 mM. Above 5 mM no increase in activity was observed. In the absence of a $CaCl_2$ a slight inhibition of activity with increasing $MgCl_2$ was

observed above 0.1 mM, as noted before [8]. From these data, the conditions for formal analysis of interpretive behaviour were chosen near the endpoints of these curves: at, and well above, the point of maximum separation (0.5 mM $MgCl_2$, indicated with the dashed black line). These conditions were 25 mM $MgCl_2$ + 10 mM $CaCl_2$ (O1), 0.5 mM $MgCl_2$ + 10 mM $CaCl_2$ (O2), 25 mM $MgCl_2$ (O3), and 0.5 mM $MgCl_2$ (O4). (B) Example gels of ribozyme reacted for 0–60 min with 2.5-fold excess S-1t substrate at 37°C (lower band: unreacted ribozyme; upper band: ribozyme reacted with S-1t substrate). Summarized data from all assays is shown in Table 1.

As seen previously, reaction progress as a function of time showed a rapid increase in product initially, with a plateau after approximately one hour. We measured the initial slope of the rapid reaction as a rough estimate of the initial velocity of the reaction to first approach the question of how well the ribozymes were performing under the various salt conditions. Given the extents of reactions of the two ribozymes under the ionic conditions chosen, we obtained the best estimates of initial velocities at 5 minutes for both ribozyme variants (see next section).

Using these initial slopes, the PV ribozyme satisfied all three inequalities needed for interpretive behaviour (O1 > O2, O1 > O3, and O4 > O2) at all time points (e.g., Figure 3b). That is, in the favourable environment the PV ribozyme is better off in the presence than the absence of the 'sign' (Ca^{2+}), and, notably, in the unfavourable environment the ribozyme is better off in the absence of the sign than in its presence, suggesting a potential evolutionary cost of misinterpretation (Table 1). We performed three replicates of each set of reactions and with the standard errors of the runs we used the Kruskal-Wallace rank-order statistic to determine that the ordering of the inequalities was marginally statistically significant for the PV at the $P < 0.1$ level.

Table 1: Percent reactivity values of the CG and PV ribozymes using the initial velocities of the pick-up-the-tail assays

Variant	O1	O2	O3	O4	O1 > O2	O1 >O3	O4>O2	O1>O4
CG	18.8	9.24	14.2	31.1	yes	yes	yes	no
PV	14.7	1.46	2.42	2.37	yes	yes	yes	yes

For both ribozymes, the initial velocities were taken using data at 5 minutes. All data are the average of three replicates. Rank-ordering of the conditions are as follows. CG: O4 > O1 > O3 > O2 (not interpretive; P = 0.019 by Kruskal-Wallace test); PV: O1 > O4 > O3 > O2 (interpretive; P = 0.075 by Kruskal-Wallace test; note that this order differs from those given from the numbers in the table for the PV because they are averages from three trials).

Lehman et al.

Lehman et al. BMC Evolutionary Biology 2014 14:248 doi: 10.1186/s12862-014-0248-2

The CG ribozyme also satisfied the three inequalities above. However, the highest performance of the CG ribozyme was in conditions of low [Mg^{2+}] and the absence of the sign (Ca^{2+}), i.e., outcome O4. By definition, for the CG ribozyme the favourable condition is therefore low rather than high [Mg^{2+}], in which case the columns of the 2×2 matrix need to be reversed in order to assess whether the relevant inequalities are satisfied. When this is done the CG ribozyme does not meet the necessary criteria for interpretive behavior. This result does not depend on our choice of time points; the same trends are observed at any time up to 15 minutes (data not shown).

For the PV ribozyme, the O3 and O4 are relatively equal values. For this ribozyme, the ordinal ranking can be described as O1 > O3 ~ O4 > O2. This is because the O3 vs. O4 ranking depends on which sets of data replicates are used, but the order between these two cells does not affect analysis of interpretive behavior. Thus for this variant, the presence of Ca^{2+} more dramatically alters the differentiation between the favorable and unfavorable states, allowing the ribozyme to better distinguish the disparity in Mg^{2+} content and hence be interpretive of its environment.

Kinetic Analyses with Radiolabeled Substrate

It was possible to estimate k_{cat} values from the assays performed with ^{32}P-labeled substrate under conditions of enzyme excess for all scenarios except for the PV in O2 and O4 where the extents of reaction were not significant even after one hour. These values are summarized in Table 2. Here it can be observed that there is general agreement with the satisfying of the inequalities that constitute minimum criteria for the possibility of interpretive behaviour. However, because two of the PV values were not determinable, and because the enzyme excess conditions do not emulate those expected to drive the evolution of this trait, we did not rely on these data for our analysis of interpretation.

Table 2: Rough estimates of k_{cat} values under conditions of enzyme excess

Variant	O1	O2	O3	O4	O1 > O2	O1 > O3	O4 > O2	O1> O4
CG	7.0×10−3	2.3 × 10−7	5.5 × 10−3	5.5 × 10−5	yes	yes	yes	yes
PV	6.9×10−3	< 10−8	1.0 × 10−3	< 10−8	yes	yes	N.D.	yes

All values given in units of min−1. For the PV, yields under the O2 and O4 conditions were too small to estimate rate constants. Rank-ordering of the conditions are as follows. CG: O1 > O3 > O4 > O2; PV: O1 > O3 > O4 & O2. However, the conditions of enzyme excess are not representative of an evolutionary interpretive scenario, hence the data in Table 1 were used for this purpose. N.D. = not determined.

Lehman et al.

Lehman et al. BMC Evolutionary Biology 2014 14:248 doi: 10.1186/s12862-014-0248-2

These ^{32}P-labelling kinetic data were used to establish that the 5-minute time point was the optimal estimate of the initial velocities for the CG and PV ribozymes. From the extents of trans-splicing reactions of the ribozymes under the condition of enzyme excess over all four salt concentration combinations, we concluded that the CG consistently reacts approximately five times faster than the PV, and that the initial velocity values (where the extents of reaction are well differentiated yet below 50%) are best measured at 5 minutes, averaging over all four conditions (Figure 4). Moreover, the level of substrate cleavage under these assays was the highest for both ribozymes in the O1condition, which was not the case for the CG ribozyme under the pick-up-the-tail assay that tracked the behavior of the ribozyme itself. This result influenced our understanding of why the PV ribozyme is interpretive but the CG is not (see Discussion).

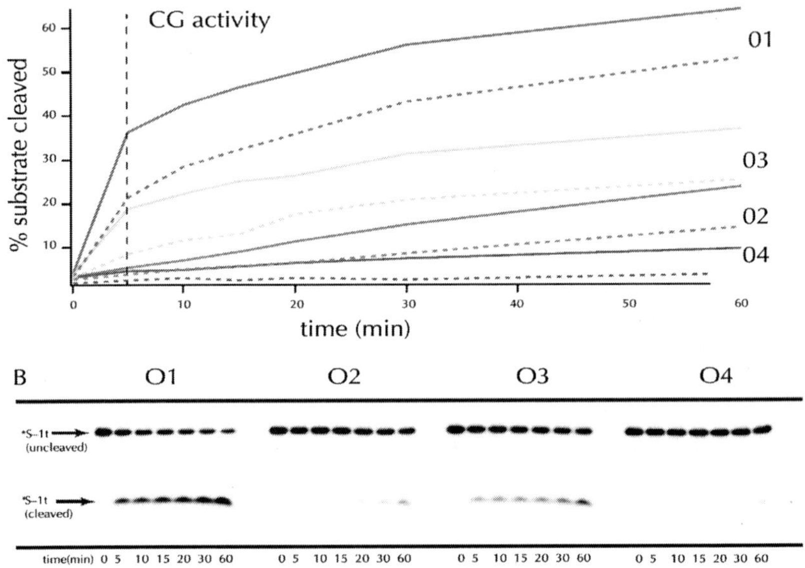

Figure 4: Semi-formal kinetic analysis of ribozyme activities under the four environmental conditions O1–O4 using 5′-32P-radiolabeled substrate-cleavage assays. (A) graph of percent substrate reacted as a function of time for the CG enzyme at two different concentrations: 25 nM (solid lines) and 10 nm (dotted lines). Horizontal dashed line indicates 5 min time point, where pick-up-the-tail data for interpretive analysis (Figure 3) was selected.(B) Example radiogram of cleavage patterns in the case of the CG ribozyme at 25 mM and *S-1t at 1 nM.

Data Analysis

An analysis of the total parameter space, and of the fitnesses of the PV ribozyme under study within that space, provides an opportunity to quantify interpretive behaviour of the ribozyme. As outlined in the Methods using equations (1), (2), and (3), the interpretive behaviour of the ribozyme can be quantified using an analysis of the total parameter space, defined by assigning values to three variables: the probability of high Mg^{2+} concentration (favourable environment), the probability of the presence of Ca^{2+} (presence of the sign) and the correlation, r, between the favourable environment and the presence of the sign. The proportion of total activity (fitness) of the ribozyme attributable to the property of interpretation is expressed by the interpretive index, I. If the total fitness of the ribozyme at a given point in parameter space is Wt, and the fitness that would obtain if there were no correlation (predictive value) between the presence of the sign and the favourable environment is W_c, then $I = (W_t - W_c)/W_t$.

Figure 5a shows that, for the PV ribozyme, there is a positive interpretive component of fitness provided that there is a correlation between the presence of the sign (Ca^{2+}) and the occurrence of the favourable environment (high concentration of Mg^{2+}). As expected, the interpretive component of fitness rises with the degree of correlation between these variables.

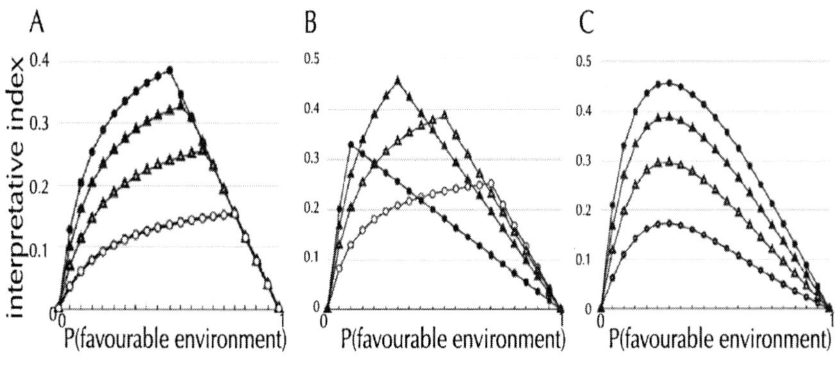

Figure 5: Analyses of the interaction between interpretation and fitness for the PV ribozyme. Data are for reaction times of 5 minutes.(A) The parameter-space in which the probability of the presence of the sign (10 mM Ca^{2+}) is 0.5 and the probability of the presence of 25 mM Mg^{2+} (as opposed to 0.5 mM Mg^{2+}) varies from 0 to 1 (x-axis). The proportion of total product attributable to the interpretive component (I) of the ribozyme activity is shown on the y-axis. Four different scenarios from the total parameter space are illustrated, each with a different correlation coefficient (r) between the probability of the sign (10 mM Ca^{2+}) and the probability of the favourable environment (25 mM Mg^{2+}): r = 1 (closed circles), r = 0.75 (closed triangles), r = 0.5 (open triangles), r = 0.25 (open circles). (B) The proportion of total activity of the PV ribozyme attributable to interpretation (y-axis) at varying probabilities of high (25 mM) Mg^{2+} (x-axis) with varying probabilities of the presence of the sign (10 mM Ca^{2+}) and maximal correlation between the sign and the favourable environment (r = 1). Probability of the presence of the sign = 0.1 (closed circles), 0.3 (closed triangles), 0.5 (open triangles), 0.7 (open circles). For each of these curves a further set would be generated by lower levels of

correlation, corresponding to the nest of curves in panel A. (C) Maximum available interpretive benefit (y-axis) at varying probabilities (x-axis) of the favourable environment (25 mM Mg^{2+}) at various correlations (r) between the sign and the favourable environment: r = 1 (closed circles), r = 0.75 (closed triangles), r = 0.5 (open triangles), r = 0.25 (open circles). Each calculated point corresponds to a peak on a member of the set of curves of which four are shown panel B.

Figure 5b shows the effect of varying the probability of the sign (presence of 10 mM Ca^{2+}) when the correlation between the sign and the presence of the favourable environment (25 mM Mg^{2+}) is maximal. Note that for a given probability of the favourable environment the maximal interpretive benefit occurs when the probability of the favourable environment is equal to the probability of the sign. This is as expected because it is the condition under which false positive and false negative responses to the sign are minimized. At higher levels of probability of the favourable environment the interpretive benefit is limited by the occurrence of false negatives (absence of the sign in the presence of the favourable environment). At lower levels of probability of the favourable environment the interpretive benefit is limited by false positives (presence of the sign in the absence of the favourable environment). (Figure 5a shows that the maximization of interpretive benefit when the probabilities of the sign and favourable environment are equal holds at all degrees of correlation between them).

Of the three parameter spaces represented in Figure 5b the greatest interpretive benefit is available where the probability of the favourable environment is close to 0.3. Calculations (not shown) reveal that the maximum occurs at a probability of the favourable environment of 0.33. The relationship between interpretive benefit and the probability of the favourable environment is explored further in Figure 5c, which shows the maximum interpretive benefit available at different probabilities of the presence of a favourable environment (i.e., parameter spaces in which the probability of the sign is equal to the probability of the favourable environment). As expected, interpretive benefit increases with increasing correlation between the sign and the interpretive environment.

DISCUSSION

The results of the assays above provide an empirical demonstration of interpretation as manifested at the molecular level. In addition, because the specific inequality ratios vary from one RNA sequence to another, we can infer that interpretation is an evolvable trait. When using the term 'interpretation' about a molecular property we are not suggesting any form of consciousness or deliberate choice; at the same time, from these results we do claim that a property more familiar in the behaviour of organisms is both definable and demonstrable at the molecular level [3].

We have shown that the PV ribozyme meets the necessary criteria for interpretive behavior. The maximum interpretive benefit (interpretive index) occurs at locations in parameter space at which the probabilities of the sign (presence of 10 mM Ca^{2+}) and the presence of the favourable environment (25 mM Mg^{2+}) are equal. This reflects the conditions under which the effects of false positive and false negative responses to the sign are minimized. The maximum interpretive benefit for the PV ribozyme at reaction times of 5 minutes occurs at the location in parameter space at which the probability of the favourable environment is 0.33. When the correlation between sign and favourable environment is maximal, and the probabilities of the sign and favourable environment are equal, the interpretive component of total fitness at this location in parameter space is 46% (Figure 5C). Even when the correlation is relatively low, r = 0.25, the interpretive component of total fitness is 17%. Levels of interpretive benefit sufficient to be subject to selection are also seen in less optimal areas of parameter space, in which the probabilities of the sign and favourable environment are not equal (Figure 5B). Importantly, the characteristics of the relationship between interpretive benefit and the parameter space variables illustrated in Figure 5 are properties of the whole system; i.e., they are 'emergent' properties of the 2 × 2 outcome matrix within the constraints defined by the values assigned to the parameter space variables. Therefore, although at a molecular level of description the physico-chemical interactions of a ribozyme molecule are adequate to explain its properties, a fully adequate explanation of a proto-biotic ribozyme sequence and structure might require reference to selection for interpretive behaviour, such an explanation necessarily involving reference to the ribozyme's past or present relation to a probabilistically varying environment. We emphasize, however, that the PV variant of the Tetrahymena ribozyme represents merely a proof-of-principle in this regard: we are not suggesting that this particular ribozyme had any proto-biotic role.

In the absence of an analysis in terms of interpretation, the values observed for the artificially selected PV version of the Tetrahymena ribozyme in the payoff matrices would be somewhat surprising, because they show that the ribozyme does better in 0.5 mM $MgCl_2$ (O4) than in the same situation but with 10 mM $CaCl_2$ present (O2). This stands in contrast to the comparison between O1and O3, where it is seen that the addition of $CaCl_2$ to a 25 mM $MgCl_2$ buffer is advantageous. From a molecular perspective this can be explained through the realization that high amounts of a poorer cofactor (Ca^{2+}) solution can flood the catalytic site and out-compete the preferred cation [8]. Thus the results seen here cannot be explained by a simple cooperative binding effect of one metal ion on another. Similarly, for the CG ribozyme, its low activity in O1 can be attributed to the tendency of theTetrahymena ribozyme to become trapped in unfavorable folded conformations at high, non-physiological concentrations of Mg^{2+} values [8],[14]. This can be detected in the pick-up-the-tail assays where the ribozyme is being tracked but is not seen in the kinetic analyses of ^{32}P-labeled substrate cleavage where there is vast enzyme excess (Figure 4).

But importantly, from the perspective of evolutionary fitness, the concept of interpretation adds a level of understanding of why certain patterns of cation utilization might be under selection. The concept of interpretation is therefore a significant and novel way of viewing evolutionary options in molecular systems. By considering that the sign (Ca^{2+}) and the object (Mg^{2+}) typically co-occur in biotic environments and in fact the former is typically at a concentration that is an order of magnitude higher than the latter (Table 3), an appreciation of a interpretive entity's ability to exploit a chemically less proficient resource – and of the evolutionary advantage this brings – allows us to understand seemingly maladapted traits (e.g., here, O4 > O2). This would be true in other molecular (and organismal) systems as well. It is worth emphasizing that we are not simply re-expressing the behaviour of the Tetrahymena chemical system in new language, but rather that we are positing that interpretation is a general evolutionary strategy operative from the molecular level upwards.

Table 3: Comparison of the physical and geological characteristics of the two divalent cations under study as putative signs in this paper

Chemical properties					Modern conditions		Early life conditions		
Cation	Hydration number	Ionic radius (Å)	pKa	Hardness (c)	Oceans (mM)	Cells (mM)	Archaean ocean (%)	Hydrothermal vents (mM)	Affinity for montmorillonite clay (ka/kCd)
Mg2+	6	0.72	11.42	47.59	50	30	24	51	0.9
Ca2+	8	1.12	12.70	19.52	10	1	29	232	5.8

Data taken from references [15]-[18].

Lehman et al.

Lehman et al. BMC Evolutionary Biology 2014 14:248 doi: 10.1186/s12862-014-0248-2

The evolutionary scenario that we envisage from these results describes at least three distinct temporal stages (Figure 6). The first stage would be that of primordial, nearly random, and simple RNA sequences not optimized for replicative activity by natural selection. These RNAs would take advantage of the dominant divalent cation in their environment (Ca^{2+}) to facilitate catalysis [19]. The binding would have been weak and many RNA sequences would have been able to conform to the binding of the relatively large and diffuse Ca^{2+} ion; but any anionic backbone shielding and/or catalytic prowess provided by this ion would have been potentially beneficial. Here the Ca^{2+} ion would have been the object itself, and not a sign.

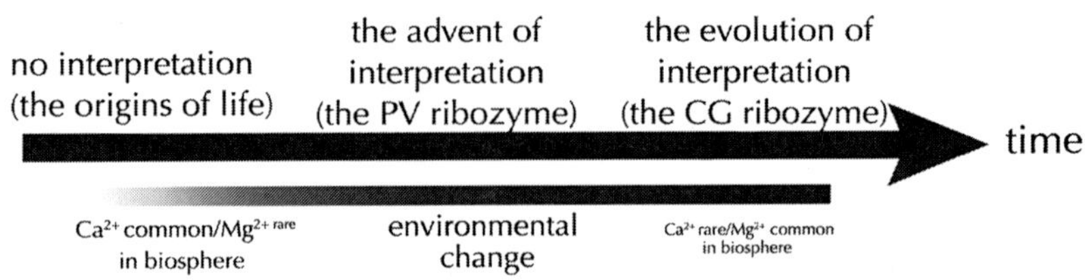

Figure 6: Proposed evolutionary progression of interpretation in theTetrahymenaribozyme system. At or near the origins of life on the Earth 4 Ga, there were no catalytic RNAs with any interpretive ability. Selection for interpretation in a relatively calcium-rich but magnesium poor environment (Table 3) drove the advent of RNAs akin to the PV, which used Ca^{2+} as a sign for the preferred Mg^{2+} ion. As the abiotic and biotic environments evolved to sequester Mg^{2+} + ions in cells, the advantage to interpretation waned, and in contemporary cells relatively enriched in Mg^{2+} + RNAs such as the CG evolved under pressures to maximize their use of Mg^{2+}.

The second stage would have been the advent of ribozymes with interpretive behavior, and the PV is seen as offering a proof-of-principle analogy for the actual RNAs that could have evolved during this time period. Here, the advantage of using Ca^{2+} as a sign in environment where the presence of $Ca^{2+}$and Mg^{2+} was correlated would imbue such sequences with a fitness advantage as they could accommodate both ions but make use of the chemically more powerful Mg^{2+} ion when available. More specifically, the structural complexity of the RNA had reached a point where the number and conformation of its cation biding sites allowed for a simultaneous distinction between, yet rudimentary simultaneous usage of, both Ca^{2+} and Mg^{2+}. The ribozyme can use the former as a sign for the latter, and we expect that simulation models will later show that this feature is enhanced in environments where the appearance of these two ions is correlated in time and space. At this evolutionary stage interpretive molecular behaviour would be advantageous and would be under positive selection.

The third stage would have been the evolution of highly specific ribozyme sequences (and their corresponding 3° structures) able to form highly specific and tight active sites that could make maximal use of the preferred Mg^{2+} ion. These sequences (e.g., the CG variant that is extant in biology today) would have the benefit of long periods of evolutionary optimization in a defined cellular environment where ion transporters could use energy to pump ions against an environmental concentration gradient and create an optimal internal milieu, such as one that is enriched in Mg^{2+}compared to Ca^{2+}. Here the evolutionary pressure for this particular kind of molecule would be for maximal and controllable catalytic activity, rather than for the ability to take advantage of signs of the state of the external environment, and thus interpretive behaviour would have been diminished or lost as it would no longer confer any selective advantage. Of course the use of signs (more often referred to as messages or signals in biological contexts) remains an important aspect of intracellular and inter-cellular communication. Our point is that in certain circumstances there may be a selective benefit to interpretive behaviour at one stage of evolution which is then superseded by a more efficient non-interpretive strategy based on high specificity binding. It is even conceivable that biotic systems ultimately retained the use of primordial signs such as Ca^{2+} for intercellular messaging purposes where the ion (sign) now has an arbitrary relation to object (intracellular or extracellular state) that it signifies.

It may be the case that other biopolymers such as proteins could have experienced evolutionary pressures based on interpretive abilities. Yet the potential central role of RNA in early life as a consequence of its dual genotypic and phenotypic functionalities makes it an ideal candidate for searching for interpretive behaviour. We anticipate that this characteristic could be evident in other RNA systems, particularly RNAs with known alternative folded structures, such as riboswitches, as discussed previously [20]. It may also be the case that other signs and objects could be utilized by RNA in an interpretive fashion. The divalent cations Mg^{2+} and Ca^{2+} are both alkali earth metals with similar, but not identical ionic radii (Table 3), and thus display many of the same chemical properties such as aqueous solubility and mineralization with subtle but often critical differences. This made them likewise good candidates for sign/object partners, and the well-studied metal-dependence [8],[10],[12] of the Tetrahymena ribozyme made it a logical system in which to investigate interpretive behaviour. Other ions such as Mn^{2+}[8] and Fe^{2+}[21],[22] are also of interest in this regard, given the roles they can play in RNA-based catalysis.

In a system in which the definition of interpretation and the three inequalities discussed above are all satisfied, it is worth considering how such an interpreting entity might evolve further. Such further evolution might include: response to a different sign that is more highly correlated with the 'object' (the entity of which the sign is functioning as a sign); greater specificity of binding of the sign (reducing errors involving analogues of the sign); strategies for reducing the disadvantage of binding the sign when the object is not present; or, in some cases, enhanced capacity to recognize the object, such that recognition of the sign is no longer advantageous. (The latter scenario is the one that we envisage could explain the evolution of the current genotype, CG, of the Tetrahymena ribozyme from the primitive variant, PV.) We propose that it should be possible to observe each of these possibilities by performing targeted artificial evolution experiments on ribozymes in which interpretation has been demonstrated. From the last possibility it may be postulated that interpretations at the molecular level may be evolutionarily more primitive than high-specificity binding, and may have been more widespread than the latter in the early stages of the evolution of life.

CONCLUSIONS

Using variants of the Tetrahymena ribozyme, we have shown that they can meet the minimum criteria for the evolution of interpretive behaviour in regards to their responses to divalent metal ion concentrations in their environment. In particular, one variant is interpretive, while another (the contemporary sequence) is not, suggesting that an evolutionary transition can take place and that interpretiveness can be under the influence of selection, either directly in a highly variable or unpredictable environment, or indirectly as in a laboratory selection for a specific phenotype. Interpretation must take place through a physical-chemical interaction between larger genotype molecules and smaller sign molecules; most likely binding constants at catalytic, general folding, or specific allosteric sites on genotypic polymers will prove to be the critical factors. Computer simulations of such events and/or additional examples of putative interpretive behavior in other RNAs will be quite valuable. Interpretation in RNA molecules provides a distinct type of intermolecular interaction event that can shape the evolutionary trajectory of macromolecules, especially in the earliest stages of life's history.

METHODS

RNA Preparation

The variants of the 393-nucleotide Tetrahymena ribozyme were transcribed in vitro from DNA templates as described previously [10]. The two ribozymes were purified by electrophoresis through 8% polyacrylamide / 8 M urea gels and resuspended in 0.1 mM EDTA at 10 pmol/μl concentration prior to use. The trans-splicing substrate for in vitro assays, the 25-mer RNA 'S-1t' [23]: 5′-GGCCCUCU•AAAUAAA UAAAUAAAUA-3′, was purchased commercially from IDT (Coralville, IA). Upon reaction with the Tetrahymena ribozyme, this molecule is spliced at the dot (.) such that the 3′ end is transferred to the 3′ end of the ribozyme and the 5′ end is liberated.

Ribozyme Assays

The initial assays were performed in a 'pick-up-the-tail' format [12],[24] in which the reaction was followed by tracking the length increase of the ribozyme following addition of the 3′ end of the substrate molecule. Successful reaction between the L–21 variants of the Tetrahymena ribozyme and the S-1t substrate results in a length change from 393 to 410 nt, which can be followed electrophoretically. For assays, 0.5 μM ribozyme was incubated with 1.25 μM substrate for 0–60 min in four different reaction buffers, all of which were buffered to pH 8.0 with 30 mM EPPS. Previously we tracked the reactivity in 0.5–50 mM $MgCl_2$ after 195 min (Figure 4 in ref. [12]), and from this we opted to perform subsequent studies for shorter times (0–15 min; to address better the ribozymes' initial velocities) in the following four reaction buffers (final concentrations): 25 mM $MgCl_2$ + 10 mM $CaCl_2$ (O1), 0.5 mM $MgCl_2$ + 10 mM $CaCl_2$ (O2), 25 mM $MgCl_2$ (O3), and 0.5 mM $MgCl_2$ (O4). These concentration values are similar to, but outside of the range of, those investigated earlier [8] for the CG ribozyme. (The existence of a Ca^{2+}-competent mutant was not known when that study [8] was performed.) Reactions were stopped by the addition of enough EDTA to chelate stoichiometrically all divalent cations in solution, and the products were electrophoresed through 8% polyacrylamide / 8 M urea gels. The resulting bands were detected and quantified after staining with SybrGreen II on a Typhoon Trio$^+$ imager and quantified using the ImageQuant software (GE Healthcare). The small but clear difference between the reactant (393 nt) and the product (410 nt) allowed the employment of SybrGreen staining with small error. All reactions were performed in triplicate.

Kinetic Studies

While the assays described in the previous section were entirely satisfactory (see Results) and performed under conditions of substrate excess, which would be pre-biotically relevant if one considers that nascent RNA genotypes would be far outnumbered by any substrate or "food" molecules, we performed additional analyses under conditions of ribozyme excess for the two ribozyme variants under differing environmental conditions following established protocols [10],[12]. Here, the reaction was followed by tracking the length loss of the substrate following transposition of its 3′ end to the ribozyme and subsequent release of the 5′ portion, which was radiolabeled with γ-^{32}P[ATP] by the use of the Optikinase enzyme (USB Biochemicals, Cleveland, OH). For assays, 10 or 25 nM ribozyme was incubated with 1.0 nM substrate for 0–60 min in the four different reaction buffers O1–O4 described above. During this period, with this stoichiometry, less than 50% (and usually less than 20%) of the substrate has been converted to product such that single turnover conditions were maintained. Reactions were stopped by the addition of enough EDTA to chelate stoichiometrically all divalent cations in solution, and the products were electrophoresed through 16% polyacrylamide / 8 M urea gels and the resulting bands were detected by phosphorimaging on a Typhoon Trio⁺ imager and quantified using the ImageQuant software (GE Healthcare). The apparent rate constants kobs were estimated by assuming that substrate binding was not rate limiting and that the reaction could be approximated with by first-order kinetics. While not strictly true for theTetrahymena ribozyme under lower salt and higher temperature conditions (10 mM $MgCl_2$ and 50°C)[14], this assumption is likely a good one averaged over all four reaction conditions studied here where the salt concentrations are typically higher and the temperature lower (37°C). The values ofk_{obs} were obtained by fitting the fraction substrate cleaved (f) data to the equation $f = A\{1-\exp(-k_{obs}t)\}$, where A is the asymptotic maximum extent of reaction. The pseudo-first-order rate constants k_{cat} were obtained estimating the y-intercept of plots of k_{obs} vs. $k_{obs}/[E]$ for each ribozyme in each condition (O1–O4). Because the vast ribozyme excess did not emulate conditions expected for interpretive scenarios, these data were primarily used to establish the best time point to assay the ribozymes for interpretive behaviour using data from the previous section (see Results), and only secondarily to assess interpretive behaviour.

Parameter-space Analysis

An analysis of the total parameter space, and of the fitnesses of the PV ribozyme under study within that space, provides an opportunity to quantify interpretive behaviour of the ribozyme. The parameter space is defined by assigning values to three variables: the probability of high Mg^{2+} concentration (favourable environment), the probability of the presence of Ca^{2+} (presence of the sign) and the correlation between the favourable environment and the presence of the sign.

The proportion of total activity (fitness) of the ribozyme attributable to the benefit of interpretation is calculated as follows. The total activity (fitness) of a ribozyme (Wt) at any location within this parameter space is given by the sum of the four experimentally measured outcomes in the 2 × 2 matrix (O1, O2, O3, and O4), weighted according to the probability of each outcome, p(O1), p(O2), p(O3), p(O4):

$$W_t = \mathbf{O1}\cdot p(\mathbf{O1}) + \mathbf{O2}\cdot p(\mathbf{O2}) + \mathbf{O3}\cdot p(\mathbf{O3}) + \mathbf{O4}\cdot p(\mathbf{O4}) \qquad (1)$$

The component of total fitness attributable to interpretation at any location in the parameter space (the interpretive index) can be quantified by correcting for the activity calculated for the ribozyme expected in the absence of any interpretive benefit. This 'control' fitness (Wc) is given by the total fitness that would obtain if there were, in fact, no correlation between the presence of the sign and the favourable environment: i.e., if the probabilities of these were independent. If the probability of the sign is p(S) and the probability of the favourable environment is p(E), then:

$$W_c = \mathbf{O1}\cdot p(S)p(E) + \mathbf{O2}\cdot p(S)(1-p(E)) + \mathbf{O3}\cdot(1-p(S))p(E) + \mathbf{O4}\cdot(1-p(S))(1-p(E)) \qquad (2)$$

When the correlation, r, between the presence of the sign and the presence of the favourable environment is zero, the probability of their co-occurrence, p(O1), is given by p(S)•p(E). In that caseWt = Wc and the presence of the sign carries no predictive value with respect to the presence of the favourable environment. When the correlation is maximal, r = 1, the probability of the co-occurrence of sign and favourable environment, p(O1) is equal to the lower of p(S) and p(E). Between this minimum and maximum, the probability of co-occurrence of sign and favourable environment is linearly related to the correlation coefficient, r. For any value of p(O1) within parameter space, the values of p(O2), p(O3) and p(O4) are determined by the fact that p(S) = p(O1) + p(O2) and p(E) = p(O1) + p(O3).

The proportion of total fitness attributable to interpretive benefit (the interpretive index, I) is then given by:

$$I = (W_t - W_c) / W_t \qquad (3)$$

AUTHORS' CONTRIBUTIONS

NL, TB, and BL performed the RNA preparations and assays described in this study. AR and CS developed the concept of interpretation and proposed its implementation in an origins-of-life context. NL conceived of the study, participated in its design and coordination, and drafted the manuscript. AR and CS helped to write the manuscript. All authors read and approved the final manuscript.

ACKNOWLEDGEMENTS

We would like to thank N. Vaidya, J.A.M. Yeates, and D. Gofreed for help during preparation of this manuscript. This work was supported by a STARS grant from the Templeton Foundation to C.S., A.R., and N.L., the Oregon Space Grant Consortium, and the Center for Life in Extreme Environments at Portland State University.

REFERENCES

1. Joyce GF: RNA evolution and the origins of life. Nature 1989, 338:217-224.
2. Pross A: Toward a general theory of evolution: extending Darwinian theory to inanimate matter. J Syst Chem 2011, 2:1.
3. Larson BC, Jensen RP, Lehman N: The chemical origin of behavior in rooted in abiogenesis. Life 2012, 2:313-322.
4. Emmeche C, Kull K: Towards a Semiotic Biology: Life is the Action of Signs. Imperial College Press, London, UK; 2011.
5. Robinson A, Southgate C: A general definition of interpretation and its application to origin of life research. Biol Philos 2010, 25:163-181.
6. Leong TL, Yang ZR, Robinson AJN, Southgate CCB: Interpretation and the origin of life. Biol Theory 2010, 5:112-116.
7. Bernhardt HS: The RNA world hypothesis: the worst theory of the early evolution of life (except for all the others). Biol Direct 2012, 7:23.

8. Grosshans CA, Cech TR: Metal ion requirements for sequence-specific endoribonuclease activity of the Tetrahymena ribozyme. Biochemistry 1989, 28:6888-6894.

9. Pan T, Long DM, Uhlenbeck OC: Divalent Metal Ions in RNA Folding and Catalysis. InThe RNA World, First Edition. Edited by Gesteland RF, Atkins JF. Cold Spring Harbor Laboratories Press, Cold Spring Harbor, NY; 1993:271-302.

10. Cernak P, Madix RA, Kuo LY, Lehman N: Accomodation of Ca(II) ions for catalytic activity by a group I ribozyme. J Inorg Biochem 2008, 102:1495-1506.

11. Kruger K, Grabowski PJ, Zaug AJ, Sands J, Gottschling DE, Cech TR: Self-splicing RNA: autoexcision and autocyclization of the ribosomal RNA intervening sequence of Tetrahymena. Cell 1982, 31:147-157.

12. Lehman N, Joyce GF: Evolution in vitro of an RNA enzyme with altered metal dependence. Nature 1993, 361:182-185.

13. Guo F, Gooding AR, Cech TR: Structure of the Tetrahymena ribozyme: base triple sandwich and metal ion at the active site. Mol Cell 2004, 16:351-362.

14. Herschlag D, Cech TR: Catalysis of RNA cleavage by the Tetrahymena thermophilaribozyme. 1. Kinetic description of the reaction of an RNA substrate complementary to the active site. Biochemistry 1990, 29:10159-10171.

15. Feig AL, Uhlenbeck OC: The Role of Metal Ions in RNA Biochemistry. In The RNA World, Second Edition. Edited by Gesteland RF, Cech TR, Atkins JF. Cold Spring Harbor Laboratories Press, Cold Spring Harbor, NY; 1999:287-319.

16. Ronov AB: Probable changes in the composition of sea water during the course of geological time. Sedimentology 1967, 10:25-43.

17. de Ronde CEJ, Channer DMD, Faure K, Bray CJ, Spooner ETC: Fluid chemistry of Archean seafloor hydrothermal vents: implications for the composition of circa 3.2 Ga seawater. Geochim Cosmochem Acta 1997, 61:4025-4042.

18. Farrah H, Hatton D, Pickering WF: The affinity of metal ions for clay surfaces. Chem Geol 1980, 28:55-68.

19. Yarus M: How many catalytic RNAs? Ions and the Cheshire cat conjecture. FASEB J 1993, 7:31-39.
20. Southgate C, Robinson A: Interpretation and the origin of life. Zygon 2010, 245:345-360.
21. Athavale SS, Petrov AS, Hsiao C, Watkins D, Prickett CD, Gossett JJ, Lie L, Bowman JC, O'Neill E, Bernier CR, Hud NV, Wartell RM, Harvey SC, Williams LD: RNA folding and catalysis mediated by iron(II). PLoS ONE 2012, 7:e38024.
22. Hsiao C, Chou I-C, Okafor CD, Bowman JC, O'Neill EB, Athavale SS, Petrov AS, Hud NV, Wartell RM, Harvey SC, Williams LD: RNA with iron(II) as a cofactor catalyzes electron transfer. Nat Chem 2013, 5:525-528.
23. Riley CA, Lehman N: Generalized RNA-directed recombination of RNA. Chem Biol 2003, 10:1233-1243.
24. Beaudry AA, Joyce GF: Directed evolution of an RNA enzyme. Science 1992, 257:635-641.

Chapter 2

Environment Sensing and Response Mediated by ABC Transporters

Sarah E Giuliani[1], Ashley M Frank[1], Danielle M Corgliano[1], Catherine Seifert[1], Loren Hauser[2], and Frank R Collart[1]

[1]Biosciences Division, Argonne National Laboratory, Lemont, IL 60490, USA

[2]Biosciences Division, Oak Ridge National Laboratory, Oak Ridge, TN 37831, USA

ABSTRACT

Background

Transporter proteins are one of an organism's primary interfaces with the environment. The expressed set of transporters mediates cellular metabolic capabilities and influences signal transduction pathways and regulatory networks. The functional annotation of most transporters is currently limited to general classification into families. The development of capabilities to map ligands with specific transporters would improve our knowledge of the function of these proteins, improve the annotation of related genomes, and facilitate predictions for their role in cellular responses to environmental changes.

Results

To improve the utility of the functional annotation for ABC transporters, we expressed and purified the set of solute binding proteins from *Rhodopseudomonas palustris* and characterized their ligand-binding specificity. Our approach utilized ligand libraries consisting of environmental and cellular metabolic compounds, and fluorescence thermal shift based high throughput ligand binding screens. This process resulted in the identification of specific binding ligands for approximately 64% of the purified and screened proteins. The collection of binding ligands is representative of common functionalities associated with many bacterial organisms as well as specific capabilities linked to the ecological niche occupied by *R. palustris*.

Conclusions

The functional screen identified specific ligands that bound to ABC transporter periplasmic binding subunits from *R. palustris*. These assignments provide unique insight for the metabolic capabilities of this organism and are consistent with the ecological niche of strain isolation. This functional insight can be used to improve the annotation of related organisms and provides a route to evaluate the evolution of this important and diverse group of transporter proteins.

BACKGROUND

Bacterial transport proteins have an essential role in mediating the uptake and efflux of small molecules with the environment and the efflux of large molecules to the outer surface of the cell[1]. These proteins comprise a heterogeneous group representative of their diverse functional and cellular roles. Transporter proteins are clustered into approximately 600 different families based on the transporter classification system [2] that incorporates both phylogenetic and functional information. This system effectively organizes transporters to the superfamily or family level, but provides very limited insight into the specific ligands which are transported by these proteins. This lack of specific functional information limits our ability to link cellular metabolic capabilities with environmental signaling molecules or nutrients and generate predictive models for cellular response to environmental changes. Certainly, improved methods for functional characterization of ligands associated with the genomic set of transporters (transportome, [3]) would provide critical insight into cellular capabilities for utilizing environmental nutrients and extruding toxic compounds.

To evaluate the impact of improved functional annotation of transporter proteins, we applied a high throughput screening method for identification of protein-ligand interactions to map ligands with transporter proteins. As many transport-associated proteins are integral membrane proteins, we used solute binding subunits of the ABC transporter family as surrogates for determining specificity of these transporters. This approach was validated for a set of bacterial ABC transporters [4] and provided valuable insight into this biologically relevant class of proteins. The family of ABC transport systems is widely distributed in all three kingdoms of life and can transport a variety of substrates such as metals, small ions, mono- and oligosaccharides, peptides, amino acids, iron-siderophores, polyamines and vitamins. In bacteria, ABC uptake transporters typically consist of a combination of a solute binding, two integral membrane, and two ATPase subunits. ABC efflux pumps are thought to lack the traditional functionality associated with the solute binding subunit, but the associated proteins sometimes form larger complexes with membrane fusion proteins [i.e. the HlyD subfamily proteins, transporter classification 8.A.1.3.1, [2] and Outer Membrane Factors (transporter classification 1.B.17 [2]].

In bacteria, the number of ABC type transport systems is generally related to genome size with approximately 2-5% of the genome encoding components of the ABC family transporters [5]. However some soil bacteria contain a relatively high number of these transporters representing 40-70% of the all transporter proteins encoded in the genome. These are essential for the utilization of environmental nutrients and may reflect the competitive aspect of nutrient acquisition in the soil environment. To evaluate functional diversity of these systems, we examined the transport capabilities for *Rhodopseudomonas palustris*, a metabolically versatile bacterium commonly found in soils and water. This organism has the ability to produce hydrogen, fix carbon dioxide, and biodegrade organic pollutants and plant-derived material. The *R. palustris* CGA009 strain contains ~ 500 transport proteins representing most major transporter classes. Included in this set are ABC transporter clusters and individual components identified by the TransportDB [6] that comprise approximately 87 distinct functional units. The ABC transporter gene set is comprised of ~350 genes representing a significant fraction (~41%) of the genome apportioned to transport capability. This organism has a diverse metabolic repertoire and characterization of the ABC transporter capabilities would provide useful insight into the metabolic and cellular capacity to utilize environmental nutrients and to extrude toxic compounds of this and related organisms.

RESULTS

Target Protein Selection

The genome set of 105 candidate solute binding proteins (SBP's) of ABC transporters in *R. palustris* were selected for production and functional characterization by ligand screening. Most of the targets were extracted from TransportDB but the set was supplemented by our internal bioinformatic analysis of attributes such as genome context and protein sequence features. The prototypical periplasmic binding protein with a predicted periplasmic signal sequence comprised the majority of the target set. Seven targets did not have readily identifiable signal peptides or an N-terminal helical region (possible membrane anchor) but were included in the set based on either genome context suggesting the protein was part of an ABC transporter gene cluster (RPA1395, 2112, 2359, 2308, 4164) or from sequence homology predicting a periplasmic binding protein (PBP) domain (RPA4686, 1389, 3707). RPA3707 is annotated as a "nitrate transporter component nrtA" and categorized as an ATPase by TransportDB. Though it is part of an operon with the gene RPA3706, coding for a putative two-component response regulator antitermination factor NasT, it has a conserved PBP domain and thus was included in this target set. The experimental target set included several proteins having signal peptides or N-terminal helices which were predicted to be associated with efflux pumps. Three genes annotated as membrane fusion proteins (RPA0681, 1648 and 4088) are specific for efflux pumps and generally believed not to influence substrate specificity. These were included in the ligand screen on the basis of a recent study demonstrating metal-induced conformational changes in the ZneB protein which were suggested to indicate an active role of membrane fusion proteins in efflux resistance systems [7]. The addition of these targets to the study efflux pump associated proteins resulted in a total of 108 candidate binding proteins (BP's) targeted for the protein production and ligand screening protocols.

Interestingly, of the 108 candidate BP's, 21 were not clustered with an integral membrane and ATPase subunits (TransportDB) based on either proximity in genome and/or functional annotation (predicted substrate) from sequence homology. There were 72 total gene clusters having at least one representative of each ABC transporter component; of these, 9 transporters were associated with 2 SBP's, 1 was associated with 3 SBP's, 61 had one associated SBP. Four additional gene clusters were each indicated by associating one SBP with either an integral membrane or an ATPase subunit. One transporter (RPA2039) was predicted to have a fused integral membrane and solute binding subunits in a single polypeptide but was not included in the final list.

Protein Production and Characterization

One technical goal of this study was to benchmark the ability to clone, express and purify the genomic set of ABC transporter-associated solute binding proteins from *R. palustris* by applying a dual-vector expression strategy in the context of a high-throughput protein production process. The success rate for protein production steps (Fig. 1) is higher than reported for other typical genome scale productions experiments with bacterial proteins [8,9]. The working target set of 107 genes representing 99% of the total target set were those successfully amplified via PCR and annealed into a vector for heterologous expression in *E. coli*. The relatively high levels for production of soluble proteins are consistent with other studies on this family of proteins [10-12]. Overall, greatest attrition for cloned targets was associated with obtaining soluble protein; of the total target set, 97% expressed, but only 87% and 71% were soluble at small scale and large scale, respectively. All but two scaled soluble proteins were successfully purified and screened in the FTS assay. The final size of the screened set (69% of all targets or 75 proteins) is sufficient to provide a comprehensive assessment of the utility of the screening approach and provide insight into the nature of transport capabilities for this organism. Finally, ligand assignment was made to approximately 45% of the total protein set, which corresponds to 48 proteins or about 64% of the proteins assayed.

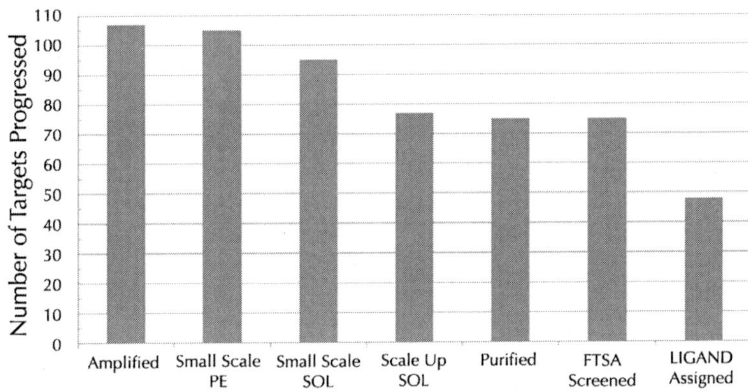

Figure 1: Experimental outcome from cloning to functional screening for the set of ABC transporter binding proteins from *R. palustris*.

Development of Ligand Library

The diversity of environmental and metabolic small molecule compounds is exemplified by the 16,000 biological compounds reported in KEGG database of known mapped metabolic pathways and cellular processes [13]. It is not currently possible to screen the complete set of candidate small molecule libraries but several strategies were invoked to enhance the ability to identify cognate ligands for the set of *R. palustris* binding proteins. One approach targeted common classes for all organisms (metals, vitamins, amino acids, peptides, and polyamines) and incorporated ligands known to bind this type of proteins. Another strategy examined the genome context of the transporter genes as these genes often cluster with known enzymes and could give an indication of metabolic capacity linked to transport. For example, select aromatic acid compounds such as benzoic acid, 4-hydroxybenzoic acid and p-coumaric acid were added to the SBP library, given knowledge of the ability of *R. palustris* to degrade aromatic acid compounds and the presence of a suite of genes involved in the degradation of benzoate and hydroxybenzoate (*3*, *4*). Results from initial binding outcomes of preliminary screens revealed the need to expand this class of ligands to probe binding specificity for similar compounds. A scan of the literature identified other similar aromatic acid compounds derived from plant lignin synthesis or breakdown such as m-coumaric acid, ferulic acid, cinnamic acid, syringic acid, quinic acid, benzylformic acid, and acetosyringone. Similarly, in response to assay detection of proteins' affinity to benzoic acid, a set of related aromatic compounds was added, including salicylic acid, benzamide, benzylaldehyde, methylvanillate, phenylacetate, and others (*5*, *6*).

Clearly, various practical limitations prevent screening of all desired ligands, such as reagent unavailability, insolubility, toxicity, expensiveness, or incompatibility with the fluorescent dye in the assay. Despite this, an attempt was made to include at least one representative ligand for various categories of chemicals for which there is a known physiological or metabolic function in bacteria or which occurs naturally in the soil or freshwater environment of *R. palustris*. Ligand categories in the library were delineated by grouping compounds with similar chemical structure, resulting in about 33 distinct categories.

Mapping Ligand and Binding Proteins in R. Palustris

The fluorescence-based thermal shift assay was used in a microwell plate format to identify candidate binding ligands for a majority of the 108 ABC transporter candidate binding proteins (Table 1). Although a binding functionality is often inferred by the NCBI annotation, the screen was conducted in an unbiased manner in that each protein was screened against the bulk of the ligand library irrespective of the inferred function. In our analysis, a chemical was considered a "binding ligand" if the calculated T_m (relative to the T_m without ligand) was 2° C or greater. The 2° C T_m significance threshold is based on previous studies [4] for this family of proteins as well as a qualitative assessment of the observed variation for proteins between the various assays (individual ligand screens, pooled screens, and replicates). Multiple binding ligands were observed for some proteins and in most cases these ligands exhibited similar chemical properties such that a general ligand binding category could be assigned. For summation purposes, a maximum of four ligands (selected on the basis of highest T_m) is listed for each target in Table 1 with the complete binding data available as additional material.

Ligand binding profiles for various proteins screened in the FTS assay were linked to additional evidence which corroborated the functional assignments, including genome context, literature descriptions for potential transport roles in metabolic pathways, and sequence homology with known PDB structures of ligand-bound proteins. A "High" data verification assignment indicates the FTSA ligand binding result is supported by good quality data and by two or more external sources of functional validation (literature, genome context, annotation/description, PDB homology, etc.). A "Moderate" data verification assignment indicates the FTSA ligand binding result is supported by one external source of functional validation while a "Low" data verification assignment indicates simply high confidence FTSA ligand-mapping data.

The outcome of this screening approach is the identification of specific roles for transporter SBP's. The spectrum of ligands mapped to the transporter binding proteins include common functionalities associated with many bacterial organisms as well as specific capabilities linked to the ecological niche that provide insight into the extraordinary metabolic diversity of *R. palustris*. The experimental results are reviewed as a summary of ligand-binding events observed for general ligand categories followed by an evaluation of the biological relevance for specific assignments.

Metal Binding Proteins

Metal binding was observed for eight proteins in the FTSA screen. This group represents a heterogeneous mix annotated as solute binding (RPA0860, RPA0884, RPA1385, RPA2410, and RPA4686), efflux pump associated (RPA0681 and RPA4088), or hypothetical (RPA4236) proteins. One protein in this cluster (RPA0860) was predicted as a metal binding protein by TransportDB (Table 1). For the most part, the predicted annotation was either nonspecific or inconsistent with the outcome of the FTSA screens. Two proteins (RPA0860 and RPA4686) were stabilized by copper, nickel, or zinc although the degree of stabilization was different. The RPA0860 protein shared domain conservation with the TroA superfamily of periplasmic metal binding proteins that share a distinct fold and metal binding properties. This protein is homologous (e-value = 1×10^{-35}) to a predicted periplasmic Zn^{2+} binding protein PsaA from *Streptococcus pneumoniae*. The crystal structure of this protein has been determined with bound zinc [14,15] and alignment of the RPA0860 protein sequence to this protein demonstrates good conservation of the residues for metal binding. Gene RPA4686 is not part of an ABC transporter operon, but its protein exhibits a similar metal binding profile as observed for the RPA0860 protein. The RPA4686 protein shares sequence similarity to several categories of periplasmic binding proteins but does not contain a recognizable periplasmic signal sequence. Although we observed specific stabilization of the protein with metal ions, there is little corroborative evidence to support the metal binding function or the association of this protein with a specific ABC transporter cluster.

The RPA0884 protein exhibited specific stabilization in the presence of Fe(III)/citrate and is a part of an ABC transporter cluster. The genomic region adjacent to this transporter cluster contains several genes involved in the utilization of iron. The RPA0875 gene is annotated as a ferrocheletase, an enzyme that is present in most prokaryotes which catalyzes the insertion of ferrous iron into protoporphyrin IX to form protoheme IX (heme) [16]. The gene encoding the RPA0891 protein is annotated as the ferredoxin-containing subunit of a glutamate synthase complex.

The RPA1385 protein, which was incorrectly annotated as being a putative phosphonate transport system SBP, was shown to have a high affinity for binding vanadate. *R. palustris* is a nitrogen fixing bacteria and has been shown to utilize a vanadium nitrogenase (V-nitrogenase) as a metabolic alternative when molybdenum is limited in the environment [17]. The vanadium nitrogenase of *Azotobacter vinelandii* has been reported to catalyze both CO and N_2 reductions and thus may provide a potential link between the evolution of carbon and nitrogen cycles [18]. Prior to this research, the cyanobacterium *Anabaena variabilis* (which also contains a V-nitrogenase) was the only organism know to contain a defined high-affinity vanadate transport system [18]. In*R. palustris*, genes RPA1381-1386 are annotated as components of a vanadate nitrogen fixation system based on homology to other similar proteins. However, in *R. palustris*, initial homology search approaches were unsuccessful in attempts to identify the high-affinity vanadate transport system. Subsequent annotation efforts have proposed vanadate transport for this system as inferred from substantial homology to ABC transporter genes in the same cluster in *A. vinelandii*. However, this update has not been successfully reflected in the current NCBI or JGI annotation lists. Our ligand mapping approach experimentally identifies the RPA1385 protein as the vanadate SBP gene for this ABC transport system. This finding not only identifies a key component of the vanadate nitrogenase fixation pathway for this organism, but may also confirm a proposed hypothesis for the presence of this system in *R. palustris* which suggests that vanadate transport systems have evolved at least twice from dissimilar ancestral genes [19].

Several other screened proteins were identified as metal binding proteins but little independent experimental evidence is available to support the functional assignments. The proteins encoded by the RPA2410 and RPA4236 genes exhibited stabilization by Cu^{+2} and Zn^{+2}, respectively. Both proteins are part of a transporter cluster but there is little independent experimental evidence to support the functional assignments. The RPA0681 and RPA4088 genes are members of the HlyD family (membrane fusion protein, MFP superfamily) and are annotated as efflux pump components that link the ABC transporter in the plasma membrane with a pore in the outer membrane. Other MFP subfamilies specifically interact with other efflux pumps families such as the major facilitator superfamily (MFS) and resistance/nodulation/cell division (RND) family.

Binding Proteins for Aromatic Compounds

Six proteins demonstrated binding to aromatic compounds as their primary ligand interaction. This activity was observed for five SBP's (RPA0668, RPA0985, RPA1789, RPA4029, and RPA4648) and an efflux pump associated protein (RPA3790). Binding profiles of the SBP's group further segregated this activity on the basis of proteins that bound benzene compounds with a single carboxyl group (RPA0668,

RPA0985, and RPA4029) verses two proteins (RPA1789 and RPA4648) that bound benzene compounds with a propenoid side chain (3 carbons) rather than a single carboxyl group. The ligand profiles indicated that specificity was based on similar chemical structures for lignin degradation products such as benzoic acid and p-coumaric acid (Figure 2). In particular, RPA0668 (*hbaE*) displayed high affinity binding (T_m = 17 °C) to benzoic acid and closely related derivatives, 4-hydroxybenzoic acid, salicylate, and benzaldehyde. The gene for this SBP is clustered with ABC transporter genes (*hbaEFGHI*) and localized between two well-known operons (*bad* and *hba*) for enzymes which are involved in the initiation of benzoic acid and 4-hydroxybenzoic acid anaerobic degradation via CoA ligation [20]. The FTS assay data is the first experimental validation demonstrating the involvement of this ABC transporter, via its associated SBP specificity, in the uptake and metabolism of benzoic acid and other aromatics.

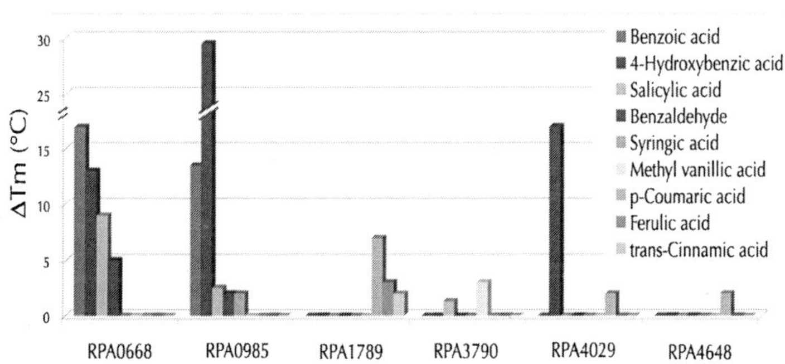

Figure 2: Identification of solute-binding proteins in *R. palustris*which bind aromatic compounds. A graph of the T_m shifts between protein with and without ligand displays six Solute-Binding Proteins which showed significant and specific stabilization with various structurally related aromatic compounds known as breakdown products of lignin in the soil environment.

Several other proteins exhibited specific binding of aromatic ligands and in several cases the ligand binding profiles were consistent with metabolic capabilities inferred from the *R. palustris*genome sequence. This organism contains several gene clusters implicated in the biodegradation of aromatic compounds [21]. Most notable are genes annotated to be involved in protocatechuate degradation (RPA4695-RPA4703, phenylacetate degradation (RPA1723, RPA1724, and RPA3765-3768), homoprotocatechuate degradation (RPA3755-RPA3762), and homogentisate degradation (RPA4670-RPA4675). The catalytic specificity of these enzymes has not been experimentally verified, but the metabolic capability generally overlaps with the observed transport profile.

Most notably, two SBP's, RPA0985 and RPA4029, exhibited very high stabilization with 4-hydroxybenzoic acid having T_m's of 29.5 and 17 °C respectively. Comparison of these two sequences using ClustalW revealed an overall significant identity (43%) and similarity (21%) of globally aligned residues. This is contrasted with alignments of RPA0985 and each of the other proteins in this group, where percent identity was less than 25%. In addition, alignment percent identity values displayed a significant positive correlation (R^2 = 0.69) to the average T_m shift for the primary shared ligand between RPA0985 and each of the other five proteins. This suggests that there are homologous residues specific to ligand binding which discriminate even between ligands with similar structures (eg. benzoate and 4-hydroxybenzoate). Further structural studies are needed to differentiate between those residues specific for ligand binding and the general sequence signatures shared by periplasmic solute binding proteins. Overall, the FTS assay seems to be a good screening tool for determining relative affinities of a protein to similar ligands as well as comparing similar proteins with the same ligand, as demonstrated with this aromatic ligand-binding set of proteins.

Additionally, one protein bound p-coumaric acid, ferulic acid, and cinnamic acid with good affinities (Figure 2). The gene encoding this SBP (RPA1789) is located on the opposite strand but near an ABC transporter operon containing 3 genes: one containing an integral membrane subunit (RPA1793), and one containing an ATPase subunit (RPA1791) , and one containing fused integral membrane and ATPase subunits (RPA1792). Two genes which are in close proximity and on the same strand as the SBP encode the enzymes p-coumaric acid-CoA ligase (RPA1787) and p-coumaroyl hydratase/lyase (RPA1786). These enzymes have been predicted to catalyze the first two catabolic steps of p-coumaric acid degradation [22]. Previously, microarray transcriptome profiling and quantitative proteomics measurements were performed with *R. palustris* cells grown on p-coumaric acid, benzoic acid or succinic acid as the sole carbon source. These studies reported significantly increased protein abundances for the two enzymes and correlating increases with the ABC transporter SBP and ATPases for p-coumaric acid verses benzoic acid or succinic acid conditions. Based on this and the FTS assay data, we propose the function for this transporter is uptake of p-coumaric acid and related compounds in the environment. Additional evidence to support the function of this cluster includes a MarR transcriptional regulator (RPA1794) which shares a putative bidirectional promoter with the three gene ABC operon and may regulate the cluster; a phenylacetic acid degradation related protein (RPA1780); a 4-hydroxybenzoate hydroxylase (RPA1781) and a TrapT family transporter operon (RPA1782-4) that are known to transport carboxylates.

Results of FTSA-determined aromatic acid binding profiles for these 6 proteins are consistent with the knowledge that *R. palustris* grows well anaerobically and aerobically on cinnamic acid, ferulic acid, and 4-hydroxybenzoic acid, and anaerobically only on benzaldehyde, benzoic acid, and DL-mandelic acid, among other substrates tested [23]. Various substrates included in the FTSA ligand library but reported not to support *R. palustris* growth were ethyl-vanillic acid (methyl form in library), phenol, quinic acid, salicylic acid, shikimic acid and syringic acid. In light of this, it is interesting to note that assumed transport of a compound based on SBP/ligand interaction such as that of salicylic acid with proteins RPA0668, RPA0985 and RPA3790 may not necessarily be correlated with an organism's ability to grow on that chemical. Still, this proposed suite of six ABC transporters for uptake of aromatic compounds in *R. palustris* expands the known list of bacterial ABC transporters which interact with aromatic compounds. The only family listed in the Transporter Classification Database http://www.tcdb.org *webcite* that uptakes aromatic-type compounds is the Taurine Uptake Transporter (TauT) Family (3.A.1.17), including the aromatic sulfonate porter, SsuABC, of *Pseudomonas putida* and the phthalate uptake system, OphFGH, of *Burkholderia capacia*[24]. Other literature suggests homologous aromatic compound transport roles of branched-chain amino acid ABC transporters in aromatic pollutant-degrading bacteria, though without experimental confirmation [25]. Aromatic compounds are more commonly transported by proteins from other families which are secondary carrier-type facilitators. Most are in the major facilitator superfamily (MFS) including the Metabolite: H+ Symporter Family, the Aromatic Acid:H+ symporter (AAHS) Family, the Anion:Cation Symporter (ACS) Family, and the Putative Aromatic

Compound/Drug Exporter (ACDE) Family. Additional families are the Aromatic Acid Exporter (ArAE) Family, the Hydroxy/Aromatic Amino Acid Permease (HAAAP) Family, and the Benzoate:H+ Symporter (BenE) Family.

Lipid-binding Proteins

Many proteins exhibited stabilization by the addition of small molecule lipids as a primary function in the FTS assay. Seventeen periplasmic binding proteins shifted in the FTS assay with either medium/long-chain dicarboxylic acid's (3 targets), medium/long-chain fatty acids (2 targets) or both (13 targets) (Table 2). There are at least three possible interpretations for these ligand-protein observations: 1) the lipid binding profile represents the functional role of the protein, 2) the binding is representative of an analog or cognate ligand, or 3) the association is the consequence of a nonspecific interaction. At this stage, it is not possible to define the nature of the interaction of most of the lipid-binding proteins in this study. However, some assignments can be corroborated by additional experimental studies and other assignments provide insight for refinement of binding studies or validation by alternative experimental approaches.

In particular, three SBP's, RPA3725, RPA3724 and RPA3723, were examined more closely, since they are clustered in the genome with ABC transporter genes. RPA3725 was assayed but did not bind any ligand. RPA3724 and RPA3723 displayed different binding profiles for various fatty acids (FA's) and dicarboxylic acids (DA's). Binding profiles indicate that the two SBP's together potentially function with the transporter to uptake a broad range of substrates having similar structures. This cluster of ABC transporter genes is adjacent to the *pimFABCDE* operon, the only identified operon to encode all enzymes for -oxidation of fatty acids & dicarboxylic acids [26]. Enzymes in this metabolic pathway degrade medium chain DA's (pimelate and longer) to glutaryl CoA and acetyl CoA. The *pimFABCDE* operon genes are required for optimum anaerobic growth on DA's and are induced aerobically with C7-C14 DA's and C8 FA. The pimA gene encodes acyl CoA ligase, which catalyzes the initial step of substrate activation with CoA. PimA ligase catalytic activity was characterized to have a broad substrate range for DA's and FA's. SBP profiles for DA's correlate significantly to pimA ligase substrate preference (Fig. 3), which suggests that the likely function of this ABC transporter is broad range substrate uptake for degradation with these enzymes. This makes sense in the context of soil bacteria, since a significant source of environmental fatty acids and dicarboxylic acids is degradation products of suberin and cutin [27] which are major lipid constituents of plants.

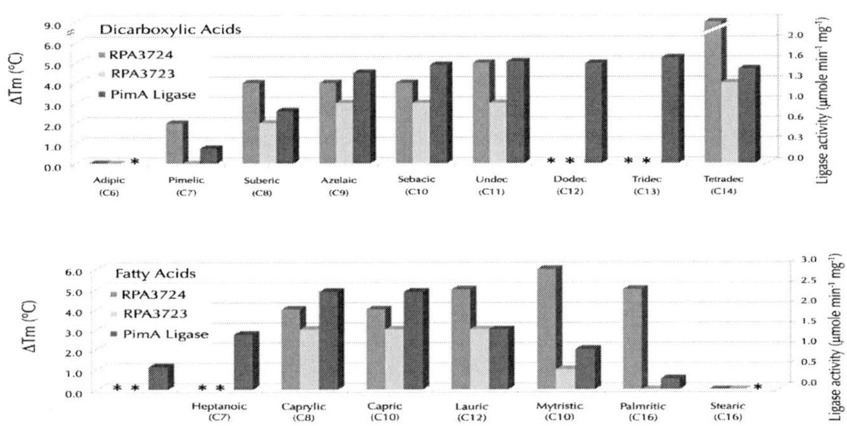

Figure 3: Comparison of binding and enzyme activity profiles for selected proteins in or adjacent to the *pimFABCDE* operon. The left ordinate represents increased stabilization increment (T_m) for the RPA3724 and RPA3723 binding proteins obtained in the presence of the indicated ligands relative to that obtained in the absence of ligands. Data represented is for dicarboxylic acid concentrations of 1000 µM assayed with 10 µM protein and fatty acid concentrations of 40 µM assayed with 4 µM protein. Theright ordinate represents the activity of PimA ligase with dicarboxylic acid and fatty acid substrates as reported in the literature[26]. **(*)** Indicates protein was not tested with the respective ligand. Name abbreviations for fatty acids and C6-C10 dicarboxylic acids lack "acid" for brevity, while abbreviations for C11-C14 dicarboxylic acids are: undec , undecanedioic acid; dodec, dodecanedioic acid; tridec, tridecanedioic acid; tetradec, tetradecanedioic acid.

The functional implication of remaining lipid-binding profiles is not immediately obvious. ABC transporters are involved in the transport and localization of lipids and their derivatives to the cytoplasm, periplasm and outer membrane as a consequence of anabolic and catabolic processes[28,29]. The transport of simple and complex phospholipids and glycolipids is mediated by ABC transporters, and these complexes are important for both the synthesis and recycling of cell wall components and membrane components [30,31]. However these lipid moieties are complex and for many outer membrane lipids the molecular form is species dependent. The current library contained only simple lipids and did not reflect the true complexity of the cellular lipid reservoir. In this context, it is reasonable to assume that some of the binding events observed in this system mimic components of the biological ligands. Delineation of the binding specificity will require expansion of the current screening library to increase representation of biologically relevant ligands and inclusion of additional experimental validation methods.

Amino Acid Binding Proteins

Three proteins (RPA2499, RPA2628, and RPA3810) were identified as amino acid binding proteins via the FTS screening (Table 1). This observation is consistent with the assigned annotation and the general functional prediction of TransportDB (Table 2). The set of binding proteins specific for amino acids provide transport capabilities for seven of the individual amino acids. The genes encoding the RPA2628 (PBP), RPA2629 and RPA2630 (integral membrane subunits) (methionine/cysteine/histidine) are orthologs of a verified amino acid transporter in *Rhizobium leguminosarum*[32]. They are also localized in a cluster of 10 ABC transporter proteins containing four periplasmic binding proteins (three screened by FTS assay,) and RPA2628 was the only protein with specific ligand binding. The characterized set of binding proteins reflects an emphasis on maintenance of sulfate and nitrogen stores. RPA2499, annotated as an arginine binding protein, is adjacent to an "amidase" gene. Asparagine has a N:C ratio of 2:4, which makes it an efficient molecule for the storage and transport of nitrogen.

Phosphate/Phosphonate Binding Proteins

Four proteins displayed stabilization with phosphate or phosphonates when screened using the FTS assay (Table 1). RPA0058 and RPA4780 exhibited specific stabilization by glycerol-3-phosphate and phosphate, respectively (Table 1). For both screened proteins, the assigned annotation and predicted substrate were in agreement and consistent with the experimental FTS assay. The RPA0699 and RPA0720 proteins had similar binding profiles and were stabilized by the addition of 2-aminoethylphosphonic acid, glycerol-3-phosphate, and phosphate. This observation is consistent with the assigned annotation for the RPA0699 protein and the general functional prediction of TransportDB (Table 2).

Urea and Polyamine Binding Proteins

Urea and Polyamine content in cells is regulated by biosynthesis, degradation and transport. In*Escherichia coli*, there are two polyamine uptake systems, namely spermidine-preferential (PotABCD) and putrescine-specific (PotFGHI), which belong to the family of ATP binding cassette transporters. Similar to the *E. coli* protein, RPA2014, predicted to bind polyamines such as putrescine, spermine and spermidine, indeed shifted with putrescine preferentially over spermine (Table 1, spermidine was not assayed). The putative associated transporter subunits are elsewhere in the genome, annotated as *potH* (RPA4159) and *potI* (RPA4158) for the two integral membrane subunits and RPA4160 for the ATPase subunit. This SBP is an example of common occurrence in bacteria where "orphan" SBP's are separated in the genome from the functionally associated transporter genes [33]. The SBP protein sequence is 51% identical to the *E. coli*putrescine binding protein (PotF) with solved structure in the PDB (1A99) and 38% identical to the*E. coli* spermidine binding protein (PotD) with solved structure in the PDB (1POT). Analysis of the respective ligands, putrescine and spermidine, in the binding sites for PotF and PotD reveal 7 key residues which confer specificity [34,35]. In PotF, these are Trp 37, Ser 38, Ser 85, Glu 185, Trp 244, Asp 247, Asp 278. All residues are identical or very similar in RPA2014 to PotF. PotF does not bind to spermidine or spermine due to two residues (Asp247 and Ser38) which affect N1 rigidity of the polyamine and are predicted to prevent the fit of longer compounds in the binding site. In PotD, one residue, corresponding to Asp247, is absent, resulting in more flexibility and space in the binding site for ligands longer than putrescine. Though these two residues (Asp247 and Ser38) are conserved in RPA2014 sequence, spermine still binds this protein with low affinity in the FTS assay. Based on these results, RPA2014 definitely plays a part in putrescine uptake, and may also transport spermine and spermidine, but further binding studies are needed to verify this observation.

Comparison and Analysis of Predicted and Experimental Annotation for R. Palustris SBPs

Despite fairly specific ligand predictions for most target proteins, only 11 of the 75 (14%) screened targets in the FTS assay exhibited binding consistent with the functional descriptions inferred from sequence homology or local genome context. These included proteins which bind urea (RPA1250, 3669), phosphate (RPA4780), sulfate (RPA0750), polyamine (RPA2014), nickel(II) (RPA2666), glycerol-3-phosphate (RPA0058), benzoate and related lignin monomers (RPA0668), di, tri, and oligo-peptides (RPA1218, RPA3691), metal cations (RPA0860, RPA0681), and phosphonic acid (RPA0699) (Table 2).

The remaining 37 targets with FTS assay ligand assignments displayed binding to a variety of compounds not predicted from sequence-based homology. Interestingly, very few of the 22 non-matching targets annotated as branched-chain amino acid or amide binding proteins (COG0683) exhibited a thermal shift with leucine, isoleucine, valine or short chain amides. Instead, these proteins bound combinations of other amino acids and/or various fairly hydrophobic compounds such as aromatic acids, medium to long-chain saturated fatty acids and/or dicarboxylic acids. Notably, 14 proteins bound ligands from more than one category and 9 of these were specific for dicarboxylic acids and fatty acids of similar structures. Only one protein, RPA3810, shifted primarily with amino acids with preferences for alanine, glycine and serine, and detectable but low affinity for leucine (Tables 2).

The other 15 targets were sorted into a variety of descriptive categories defined by COG numbers, as these designations are associated with fairly specific ligand classes for SBPs. A smaller set of 5 targets annotated as binding "nitrate, sulfate, thiamine, taurine and sulfonates" (COG0715) displayed specific affinity for compounds with multiple carboxylic acids and/or amino groups such as asparagine, malate, citrate, guanine, and thiamine. Uniquely represented in COG0687 (spermine/putrescine/polyamine) and COG3221 (phosphate/phosphonate) were RPA4648 and RPA1385 which bound p-coumaric acid and vanadate, respectively, instead of the predicted ligand classes. Also, single targets in each of the categories, COG0747 (di- or oligopeptides), COG0614 (Fe+3 siderophores, iron, cobalamine, hemin), COG0834 (lysine/arginine/ornithine), and COG1653 (sugar/maltose/trehalose, glycerol-3-phosphate), displayed affinity for long chain fatty acids, phosphates/phosphonates, metal cations, and peptides, respectively. Four proteins, assigned to either COG0845 or COG1463 as being associated with efflux of drugs, proteins, or organic toxic solvents exhibited moderate T_m shifts with aromatic compounds, zinc(II) and nickel(II).

There were 27 targets screened which did not bind any ligand in the FTS assay. Ligand categories which were present in the assay's library were indicated in the descriptions for 18 of these proteins including phosphate, iron, branched-chain amino acid, peptides, sulfonates, molybdate, and sugars or glycerol-3-phosphate. Since the ligand descriptions are fairly broad, it is likely that the specific representative library ligands in these categories are not the actual physiological ligands for these SBP's, a situation remedied by expansion of the ligand library. Finally, nine targets had only general annotation as a "extracellular ligand-binding receptor", "periplasmic solute binding" or "conserved hypothetical" protein, and are the most difficult to characterize without additional empirical data. Given the ~ 27 to 33 ligand categories present in the library, reflecting probable ligands from both predicted and experimental annotation, ~20 (66%) of these were represented in the positive functional assignments determined by the FTS assay. Predicted ligand categories likely to be identified but were not represented at all in screened target ligand profiles were nitrate and taurine/sulfonic acids.

DISCUSSION

There have been few experimental studies that characterize the transporter specificity on a genome scale and this study represents the first genome wide approach for the experimental characterization of ABC transporter proteins. The ligand screening approach identified binding ligands for 48 binding proteins associated with the set of ABC transporters. The overall ligand binding profile reflects the metabolically diversity of *R. palustris* and is consistent with characterized or inferred cellular metabolic capabilities as well as nutrient characteristics of the ecological niche. The FTS screen identified several binding proteins associated with transport of aromatic compounds and fatty and dicarboxylic acids. These capabilities are aligned with the characteristic of the isolation site, the subsurface layer of a forest litter pool, and the encoded genomic metabolic capabilities to enable utilization of structurally diverse compounds derived from degradation of plant material [21]. In many cases, these transporter complexes are co-located with clusters of genes associated with the biodegradation of aromatic compounds and fatty acids. The binding profiles for the aromatic transporters will provide a foundation for characterization of the substrate preference of the uncharacterized enzymes associated with aromatic compound degradation.

The FTS screen identified ligands and binding proteins associated with core cellular requirements of environmental organisms that reflect transport capabilities for metals, sulfate, phosphate, amino acids, peptide, and polyamines. For many of the solute binding proteins, the ligand binding assignments were supported by bioinformatic analyses or by the ability of the protein to bind chemically related ligands. Experimental observations that identified binding proteins for glycerol-3-phosphate, phosphate, sulfate, and peptides were consistent with sequence based predictions based on the original annotation or TransportDB. There was less overlap for experimental observations and sequence base predictions of metal, polyamine, vitamin, and amino acid binding proteins. The experimental screen confirmed some of the inferred binding properties but in other cases contradicted the assignment or provided a specific ligand assignment in place of a general prediction. This is not surprising in view of the limited number of binding proteins that have been characterized using biochemical or genetic methods. There are only a few categories of ligand binding proteins that have been experimentally characterized. Most of these studies examined a single or limited number of potential ligands and were not designed to examine the spectrum of the natural ligand diversity. This class of proteins also represents a challenge for specific functional annotation. Structural studies of this solute binding protein superfamily demonstrate these proteins generally exhibit a structure where two globular domains are connected by a hinge region that allows flexibility for "opened" and "closed" states [5]. The hinge region is variable in length and structure which influences the binding characteristics of the protein [36]. Particular residues in the binding sight are involved in ligand specificity, though rarely is there a large contiguous conserved sequence pattern that can be correlated to the binding site of an unknown protein without 3-D structural evidence or comparison to a homologous protein structure. Our results show many transporter solute binding proteins in *R. palustris* CGA009 are incorrectly annotated and suggest that functional characterization based on sequence analysis cannot be exclusively relied upon to predict biologically relevant and specific ligands. Given the best of predictions, there is still disparity between expected and experimentally determined function. As a case in point, the outcome for the three SBP's, RPA3723, RPA3724 and RPA3725 varied greatly. While the former two proteins bound various fatty acids and dicarboxylic acids, RPA3725 did not bind any ligand. This is surprising, since these three genes are all very closely related and form a cluster that is specific to the *Rhodopseudomonas* and *Bradyrhizobium* clade. The protein sequence for RPA3725 is 71% identical to RPA3724 and 60% identical to RPA3723. The protein-ligand interactions identified in this study will enable improvement of sequence-based predictions and provide a basis for expanding our knowledge of transporter functionality.

The results for this genome set illustrate some of the limitations of the current system for functional characterization of these proteins and indicate areas to improve the scope of functional characterization. Approximately 74% of the genome set of solute binding proteins entered the functional screening component but candidate ligands were identified for only 64% of the screened proteins. These proteins were extensively dialyzed and exhibited thermal melting suggestive of a properly folded protein structure. Although we eliminate some attrition due to loss of function or protein-ligand co-purification, the disparity between the number of screened proteins and functional assignments is largely a reflection of the limited representation of the ligand library. In addition, some of the experimentally observed interactions represent noncognate ligand binding. The selection of chemicals for the screening library was not directed towards representation of binding proteins associated with efflux transporters [37], synthesis of cellular structures [31], or recycling of cellular biomolecules [38].

CONCLUSIONS

This study provides experimental functional characterization for ABC transporter solute binding proteins from *R. palustris*. The ligand binding profiles and number of transport proteins specific for aromatic compounds is consistent with ecological and laboratory studies which demonstrate the capabilities of this organism for the utilization of plant degradation products such as lignin-derived aromatic compounds. The results of this study also provide important biological insight for the metabolic capabilities and environment fitness of this organism. This functional insight can be used to improve the annotation of related organisms and provides a route to evaluate the evolution of this important and diverse group of transporter proteins.

METHODS

Target Staging

Coding sequences for all ABC transporter associated solute-binding proteins (SBP's) were extracted from the JGI Integrated Microbial Genomes page for *Rhodopseudomonas palustris*CGA009 and the protein sequences were characterized using the SignalP[39,40] and TMHMM [41] algorithms to identify hydrophobic sequence features and guide the selection of clonable regions. The peptides specified by cloned target sequences excluded predicted N-terminal signal peptide sequences and any overlapping N-terminal trans-membrane helices. Predicted N-terminal cleavage sites were manually adjusted for input into an automated primer design tool (http://tools.bio.anl.gov:8000/bioJAVA/jsp/PeriplasmicProteinDesign/index.html *webcite*) that uses the predicted cleavage information derived from SignalP 3.0 to automatically generate amplification primers for the mature (cleaved) form of the protein.

Gene Cloning, Protein Expression and Purification

The *E. coli* cloning and expression approach employed 96-well plate-based methods and parallel expression strategies [42]. Target genes were PCR amplified from genomic DNA (ATCC:*Rhodopseudomonas palustris* CGA009 # BAA-98D-5) using a KOD HiFi DNA polymerase reaction (Novagen) combined with a touchdown PCR program designed for the GeneAmp PCR System 9700 thermocycler machine (Applied Biosystems). The 50 μl reaction was enhanced for GC-rich genomic DNA by addition of 2.5% DMSO based on previous studies [43]. Amplification products included appended 5′ and 3′ ligation independent cloning (LIC) sites to enable simultaneous cloning in multiple vectors. To optimize protein solubility outcome for predicted periplasmic proteins, an initial set of 48 genes were cloned in parallel by a LIC method [44] into a two expression vectors, pMCSG7 and pBH31. While pMCSG7 expresses protein in the cytoplasm, pBH31 is used for periplasmic protein expression. Both vectors append an N-terminal hexahistidine (6x-His) fusion tag and TEV protease cleavage recognition sequence between the fusion tag and the target protein. Additionally, pBH31 codes for an N-terminal PelB leader sequence which is cleaved off during transport through the inner membrane into the periplasm [45]. Only 8 of 48 clones from the initial set were better expressed in pBH31; thus the remainder of the SBP target set was cloned only into pMCSG7 vector. Each target was characterized for amplification, expression, and solubility using 96-well plate assays and high-density gel formats for denaturing gel analysis of proteins [46]. Targets were scored as positive for expression and solubility if a detectable fusion protein of the correct molecular weight was observed on gels stained with Coomassie-based Simply Blue Safestain (Invitrogen). Targets scored as positive for solubility were sequence verified prior to purification and screening.

Clones expressing soluble proteins were scaled to 500 ml cultures and purified using standard affinity chromatography Ni-NTA bead (Qiagen) purification methods using a combination of the automated AKTA system (GE-Healthcare) as previously described [47] and parallel manual methods. All purified proteins retained the 6x-His tag, since previous investigation of the effect of the tag on the outcome of ligand binding detection revealed insignificant tag interference for this class of proteins [4]. The purified proteins were dialyzed for buffer exchange into an assay-compatible buffer, flash frozen in liquid nitrogen [48], and stored at -80 °C until the assay. Protein concentrations were initially determined by measuring absorbance at 280 using UV-spectrophotometry then verified and assessed for purity by comparison with a BSA standard using SDS-PAGE denaturing gel electrophoresis and Simply Blue Safestain for protein visualization. Once thawed, proteins were stored at 4 °C and used within two weeks.

Fluorescence-based Thermal Shift Assay

Assay Reaction Components and Set-Up

An environmentally sensitive dye, SYPRO orange (Invitrogen, #S6650), was used at 5x concentration in all assays. Proteins were diluted to a standard concentration of either 5 or 10 μM, and screened with either 500 or 1000 μM ligand, respectively, to maintain an optimized 100x ratio of ligand to protein concentrations.

Absolute values for protein and ligand concentration did not significantly affect the outcomes of a particular reaction as long as the ligand to protein ratio was consistent. The conclusion was derived by analysis of a set of five positive control proteins with experimentally characterized values (T_m) for ligand stabilization. These targets were screened at both 5 μM protein with 500 μM ligand and 10 μM protein with 1000 μM ligand. The screening results indicated no significant difference (≤ 2 °C) in T_m shifts between reactions having different component concentrations but the same ligand to protein concentration ratio (data not shown). Test screens also indicated that this ratio is needed for optimal sensitivity for ligand binding detection. Exceptions to the standard 100x ratio were for reactions containing fatty acid ligands. These more hydrophobic compounds exhibited characteristic melt curves having fluorescence values higher than buffer background when screened with the dye at standard ligand concentrations. Reactions containing fatty acids were optimized to 40 μM ligand and 4 μM protein (10x ligand to protein concentration ratio) which decreased ligand background to minimal values, yet still enabled reasonably sensitive detection of ligand binding. Though high ligand background melt curve fluorescence was present for 500 μM or 1000 μM fatty acids, this did not always interfere with T_m shift interpretation in these reactions, especially for C8-C14 ligands. Thus, data for both 100x and 10x ligand-to-protein concentration ratios were collected and reported for targets which bound fatty acids. Accordingly, T_m shift values displayed dependence on ligand concentration and are proportionally lower for 10x ratios. All protein and SYPRO orange dilutions resulted in final assay buffer concentrations of 100 mM HEPES, 150 mM NaCl, pH 7.5 in a standard 20 μl reaction volume. Reaction volume was 40 μl only if ligands used in the reaction were soluble as concentrated stocks in 100% DMSO and not aqueous buffer. Ligands in DMSO were added for a final reaction concentration of 2% DMSO.

Assay Instrumentation, Standard Program Parameters, Data Analysis

FTS assays were performed using two different quantitative PCR instruments. Initially, the first half of the target set produced was screened using an Mx4000 multiplex quantitative PCR instrument (Stratagene) that enabled thermal manipulations and dye fluorescence detection based on a previously published method [49]. Parameters employed were the same as those previously specified for program set-up and data analysis [4]. The remaining targets were screened using the more advanced LightCycler®480 Real Time PCR System (Roche). This enabled higher density, higher-throughput data collection by acquiring 30 points per degree over a 70 °C temperature range, decreasing assay time to 20 minutes, and allowing assay setup and program execution to be independent of plate configuration. Optimized run protocol parameters were defined in a "Protein Melt" program for 1 cycle using "Melting Curve" analysis mode (Table 3). Assay reactions were performed in 96-well, white PCR plates (Axygen, #PCR-96-FLT-W or Roche, #04729692001) and wells were capped using optical 8x-strip caps (Axygen, #PCR-2CP-RT-C) or optical sealing film (Roche, #04729692001). The instrument software monitors the fluorescence in real time, creating an output protein melting curve graph of temperature (°C) vs. fluorescence. Data was displayed and analyzed using the software's "T_m calling" analysis option, which allows computation of the derivative curve displayed as temperature vs. –(d/dT) of the raw melt curve fluorescence values. From the derivative curve, the protein melting temperature midpoint (T_m) was selected as the temperature corresponding to the minimum fluorescence value. The difference in shifts in melting temperatures (T_m, °C) between a protein with and without ligand indicated a change in the protein stability; positive T_m for protein-ligand combinations was interpreted as potential ligand binding. Values for T_m's and T_m's and were calculated objectively via an automated algorithm using Microsoft Excel software and raw data exported from the instruments.

To verify that using different instruments for the complete target set minimally affected the outcomes for ligand binding detection, a set of

10 positive control proteins with known ligands and T_m's were screened with their respective ligands under identical reaction conditions in both quantitative PCR instruments (data not shown). Comparison of T_m values for protein alone indicated an average difference of 3 °C higher T_m's for reactions in the LightCycler®480 instrument versus the Mx4000 instrument. However, the difference between T_m values generated on the two instruments for reactions containing protein and ligand was less than 1 °C. This indicates a systematic increase in all values of the protein melting profiles generated by the LightCycler®480 instrument, which does not significantly affect the computed T_m values for comparable reactions with a specific target protein. Thus, the absolute T_m values are independent of the two instruments utilized for this study.

Assay Screening Method

Proteins were screened using a two-step approach: an initial screen against all pools of ligands followed by a deconvolution analysis to determine individual ligand binding. Proteins displaying positive shifts of melting temperature midpoint (T_m) with certain pools in the initial screen were screened again with the pool and also with each individual ligand present in that pool to identify specific binding ligands. Most proteins were screened against all relevant pools to provide an equal opportunity for all proteins to bind all ligands. Proteins assigned to the various COG categories 0683, 0834, 0687, 0715, which were functionally characterized prior to expansion of the ligand library, are exceptions. All reactions in which pooled or individual ligands stabilized protein were independently duplicated, and averages of the duplicate T_m values were reported. The maximum variability associated with each data point derived from averaged data of duplicate reactions was consistently less than 2 °C. In each plate experiment, negative control reactions were run for each protein with no ligand, for buffer only, and 5x SYPRO orange dye only. Fluorescence values for dye and buffer control reactions displayed no significant background thermal melting pattern compared to protein; thus, background was not subtracted from experimental fluorescence values since this correction did not affect T_m values. T_m values for all proteins were dependent on the buffer content, and for some proteins, the T_m value for a defined concentration differed significantly between reactions with and without 2% DMSO. In these cases, both T_m values were reported. All protein-only T_m's reported are an average of two of more replicates.

In the FTS assay, false negatives are expected if the native ligand is present but not detected. This can possibly result from loss of native protein structure, undetected ligand insolubility/instability, or the ligand–to-protein concentration ratio is too low to compensate for a low affinity binding or small protein/ligand complex stabilization. In the case of protein stability, 2 of 27 targets tested (RPA3961 and RPA0048) but with no ligand binding outcome did not display a clear thermal melt curve with fluorescent dye. These proteins may have been partially denatured prior to the assay, but were not repurified and retested. The remaining 25 targets were considered properly folded since a clear melt curve was reproducibly generated from fresh samples and the protein-only T_m value was consistent across replicates.

Ligand Stocks and Ligand Pools Preparation

Individual ligands were dissolved in either buffer containing 100 mM HEPES and 150 mM NaCl, pH 7.5 (1x Standard HEPES) or 100% DMSO, depending on solubility, and stored at 4°C. Exceptions were guanine and hypoxanthine, which dissolved in 1x Standard HEPES buffer at pH 10, and diaminopimelate, which dissolved in 1x Standard HEPES buffer at pH 1.5. These ligands were added to the assay so that the final amount of buffer at nonstandard pH was 2%. The cysteine stock solution contained equimolar amounts of DTT to prevent oxidation during storage and assay. All ligands were purchased from Sigma-Aldrich-Fluka-Supelco, except Putrescine (MP Biomedicals), oleic acid (EMD Biosciences), histidine (Novabiochem), cysteine (J. T. Baker); anhydrous sodium thiosulfate, D(+)-maltose, D-xylose, and iron(III) chloride (Fisher Scientific); disodium molybdate dehydrate and cupric chloride dihydrate (Mallinckrodt); and tryptone digest, anhydrous glucose and anhydrous sodium phosphate (American Bioanalytical). Ligand pools included no more than 10 ligands each and were developed systematically based on ligand chemical classification and/or compatible solubility for ease of high-throughput screening.

Ligands were considered to be stable if they were soluble in HEPES buffer or 100% DMSO at room temperature and pH 7.5. Many ligands demonstrated significant binding to more than one test protein indicating consistent, reproducible solution stability in the assay, or were previously assayed with positive control proteins (eg. amino acids). Ligands suspect of potential insolubility were those dissolved in 100% DMSO, which were added to the assay reaction as only 2% DMSO in HEPES buffer. No direct measurement was made to verify solubility in these cases except qualitative observation of precipitation or discoloration. All DMSO ligands were added to the reaction last and immediately before performing thermal denaturation to minimize insolubility. Ligand concentration in the standard reaction was 100-fold more than protein concentration to ensure reasonable detection of binding even for protein/ligand pairs with lower affinity.

AUTHORS' CONTRIBUTIONS

Conceived and designed the experiments: FC, SG, AF. Performed the experiments: SG, AF, CS, DC. Analyzed the data: FC, SG, AF. Contributed reagents/materials/analysis tools: FC LH . Wrote the paper: FC, SG, LH.

ACKNOWLEDGEMENTS

The submitted manuscript has been created by UChicago Argonne, LLC, Operator of Argonne National Laboratory ("Argonne"). Argonne, a U.S. Department of Energy Office of Science laboratory, is operated under Contract No. DE-AC02-06CH11357. The U.S. Government retains for itself, and others acting on its behalf, a paid-up nonexclusive, irrevocable worldwide license in said article to reproduce, prepare derivative works, distribute copies to the public, and perform publicly and display publicly, by or on behalf of the Government.

This article has been published as part of *BMC Genomics* Volume 12 Supplement 1, 2011: Validation methods for functional genome annotation. The full contents of the supplement are available online at

REFERENCES

1. Yen MR, Choi J, Saier MH Jr.: Bioinformatic analyses of transmembrane transport: novel software for deducing protein

phylogeny, topology, and evolution. *J Mol Microbiol Biotechnol* 2009, 17:163-176.

2. Saier MH Jr., Yen MR, Noto K, Tamang DG, Elkan C: The Transporter Classification Database: recent advances. *Nucleic Acids Res* 2009, 37:D274-278.
3. Mauchline TH, Fowler JE, East AK, Sartor AL, Zaheer R, Hosie AH, Poole PS, Finan TM:Mapping the Sinorhizobium meliloti 1021 solute-binding protein-dependent transportome. *Proc Natl Acad Sci U S A* 2006, 103:17933-17938.
4. Giuliani SE, Frank AM, Collart FR: Functional assignment of solute-binding proteins of ABC transporters using a fluorescence-based thermal shift assay. *Biochemistry* 2008, 47:13974-13984.
5. Davidson AL, Dassa E, Orelle C, Chen J: Structure, function, and evolution of bacterial ATP-binding cassette systems. *Microbiol Mol Biol Rev* 2008, 72:317-364.table of contents
6. Ren Q, Chen K, Paulsen IT: TransportDB: a comprehensive database resource for cytoplasmic membrane transport systems and outer membrane channels. *Nucleic Acids Res* 2007, 35:D274-279.
7. De Angelis F, Lee JK, O'Connell JD 3rd, Miercke LJ, Verschueren KH, Srinivasan V, Bauvois C, Govaerts C, Robbins RA, Ruysschaert JM, *et al.*: Metal-induced conformational changes in ZneB suggest an active role of membrane fusion proteins in efflux resistance systems. *Proc Natl Acad Sci U S A* 2010, 107:11038-11043.
8. Graslund S, Nordlund P, Weigelt J, Hallberg BM, Bray J, Gileadi O, Knapp S, Oppermann U, Arrowsmith C, Hui R, *et al.*: Protein production and purification. *Nat Methods* 2008, 5:135-146.
9. Savitsky P, Bray J, Cooper CD, Marsden BD, Mahajan P, Burgess-Brown NA, Gileadi O:High-throughput production of human proteins for crystallization: The SGC experience. *J Struct Biol* 2010, 172:3-13.
10. Muller A, Thomas GH, Horler R, Brannigan JA, Blagova E, Levdikov VM, Fogg MJ, Wilson KS, Wilkinson AJ: An ATP-binding cassette-type cysteine transporter in Campylobacter jejuni inferred from the structure of an extracytoplasmic solute receptor protein. *Mol Microbiol* 2005, 57:143-155.

11. Oh BH, Pandit J, Kang CH, Nikaido K, Gokcen S, Ames GF, Kim SH: Three-dimensional structures of the periplasmic lysine/arginine/ornithine-binding protein with and without a ligand. *J Biol Chem* 1993, 268:11348-11355.

12. Sun YJ, Rose J, Wang BC, Hsiao CD: The structure of glutamine-binding protein complexed with glutamine at 1.94 A resolution: comparisons with other amino acid binding proteins. *J Mol Biol* 1998, 278:219-229.

13. Kanehisa M, Goto S, Kawashima S, Okuno Y, Hattori M: The KEGG resource for deciphering the genome. *Nucleic Acids Res* 2004, 32:D277-280.

14. Lawrence MC, Pilling PA, Epa VC, Berry AM, Ogunniyi AD, Paton JC: The crystal structure of pneumococcal surface antigen PsaA reveals a metal-binding site and a novel structure for a putative ABC-type binding protein. *Structure* 1998, 6:1553-1561.

15. Loisel E, Jacquamet L, Serre L, Bauvois C, Ferrer JL, Vernet T, Di Guilmi AM, Durmort C:AdcAII, a new pneumococcal Zn-binding protein homologous with ABC transporters: biochemical and structural analysis. *J Mol Biol* 2008, 381:594-606.

16. Dailey HA, Dailey TA, Wu CK, Medlock AE, Wang KF, Rose JP, Wang BC: Ferrochelatase at the millennium: structures, mechanisms and [2Fe-2S] clusters. *Cell Mol Life Sci* 2000, 57:1909-1926.

17. Oda Y, Samanta SK, Rey FE, Wu L, Liu X, Yan T, Zhou J, Harwood CS: Functional genomic analysis of three nitrogenase isozymes in the photosynthetic bacterium Rhodopseudomonas palustris. *J Bacteriol* 2005, 187:7784-7794.

18. Lee CC, Hu Y, Ribbe MW: Vanadium nitrogenase reduces CO. *Science* 2010, 329:642.

19. Pratte BS, Thiel T: High-affinity vanadate transport system in the cyanobacterium Anabaena variabilis ATCC 29413. *J Bacteriol* 2006, 188:464-468.

20. Egland PG, Pelletier DA, Dispensa M, Gibson J, Harwood CS: A cluster of bacterial genes for anaerobic benzene ring biodegradation. *Proc Natl Acad Sci U S A* 1997, 94:6484-6489.

21. Larimer FW, Chain P, Hauser L, Lamerdin J, Malfatti S, Do L, Land ML, Pelletier DA, Beatty JT, Lang AS, *et al.*: Complete genome

sequence of the metabolically versatile photosynthetic bacterium Rhodopseudomonas palustris. *Nat Biotechnol* 2004, 22:55-61.

22. Pan C, Oda Y, Lankford PK, Zhang B, Samatova NF, Pelletier DA, Harwood CS, Hettich RL:Characterization of anaerobic catabolism of p-coumarate in Rhodopseudomonas palustris by integrating transcriptomics and quantitative proteomics. *Mol Cell Proteomics* 2008, 7:938-948.
23. Harwood CS, Gibson J: Anaerobic and aerobic metabolism of diverse aromatic compounds by the photosynthetic bacterium Rhodopseudomonas palustris. *Appl Environ Microbiol* 1988, 54:712-717.
24. Chang HK, Dennis JJ, Zylstra GJ: Involvement of two transport systems and a specific porin in the uptake of phthalate by Burkholderia spp. *J Bacteriol* 2009, 191:4671-4673.
25. Lykidis A, Perez-Pantoja D, Ledger T, Mavromatis K, Anderson IJ, Ivanova NN, Hooper SD, Lapidus A, Lucas S, Gonzalez B, Kyrpides NC: The complete multipartite genome sequence of Cupriavidus necator JMP134, a versatile pollutant degrader. *PLoS One* 2010, 5:e9729.
26. Harrison FH, Harwood CS: The pimFABCDE operon from Rhodopseudomonas palustris mediates dicarboxylic acid degradation and participates in anaerobic benzoate degradation. *Microbiology* 2005, 151:727-736.
27. Wallis JG, Browse J: Lipid biochemists salute the genome. *Plant J* 2010, 61:1092-1106.
28. Nagao K, Kimura Y, Mastuo M, Ueda K: Lipid outward translocation by ABC proteins. *FEBS Lett* 2010, 584:2717-2723.
29. Pohl A, Devaux PF, Herrmann A: Function of prokaryotic and eukaryotic ABC proteins in lipid transport. *Biochim Biophys Acta* 2005, 1733:29-52.
30. Malinverni JC, Silhavy TJ: An ABC transport system that maintains lipid asymmetry in the gram-negative outer membrane. *Proc Natl Acad Sci U S A* 2009, 106:8009-8014.
31. Sperandeo P, Deho G, Polissi A: The lipopolysaccharide transport system of Gram-negative bacteria. *Biochim Biophys Acta* 2009, 1791:594-602

32. Walshaw DL, Poole PS: The general L-amino acid permease of Rhizobium leguminosarum is an ABC uptake system that also influences efflux of solutes. *Mol Microbiol* 1996, 21:1239-1252

33. Thomas GH: Homes for the orphans: utilization of multiple substrate-binding proteins by ABC transporters. *Mol Microbiol* 2010, 75:6-9.

34. Igarashi K, Ito K, Kashiwagi K: Polyamine uptake systems in Escherichia coli. *Res Microbiol* 2001, 152:271-278.

35. Sugiyama S, Matsuo Y, Maenaka K, Vassylyev DG, Matsushima M, Kashiwagi K, Igarashi K, Morikawa K: The 1.8-A X-ray structure of the Escherichia coli PotD protein complexed with spermidine and the mechanism of polyamine binding. *Protein Sci* 1996, 5:1984-1990.

36. Berntsson RP, Smits SH, Schmitt L, Slotboom DJ, Poolman B: A structural classification of substrate-binding proteins. *FEBS Lett* 2010, 584:2606-2617.

37. Fernandez-Recio J, Walas F, Federici L, Venkatesh Pratap J, Bavro VN, Miguel RN, Mizuguchi K, Luisi B: A model of a transmembrane drug-efflux pump from Gram-negative bacteria. *FEBS Lett* 2004, 578:5-9.

38. Lee M, Zhang W, Hesek D, Noll BC, Boggess B, Mobashery S: Bacterial AmpD at the crossroads of peptidoglycan recycling and manifestation of antibiotic resistance. *J Am Chem Soc* 2009, 131:8742-8743.

39. Bendtsen JD, Nielsen H, von Heijne G, Brunak S: Improved prediction of signal peptides: SignalP 3.0. *J Mol Biol* 2004, 340:783-795.

40. Emanuelsson O, Brunak S, von Heijne G, Nielsen H: Locating proteins in the cell using TargetP, SignalP and related tools. *Nat Protoc* 2007, 2:953-971.

41. Sonnhammer EL, von Heijne G, Krogh A: A hidden Markov model for predicting transmembrane helices in protein sequences. *Proc Int Conf Intell Syst Mol Biol* 1998, 6:175-182.

42. Zerbs S, Frank AM, Collart FR: Bacterial systems for production of heterologous proteins. *Methods Enzymol* 2009, 463:149-168.

43. Choi JS, Kim JS, Joe CO, Kim S, Ha KS, Park YM: Improved cycle sequencing of GC-rich DNA template. *Exp Mol Med* 1999, 31:20-24.

44. Stols L, Gu M, Dieckman L, Raffen R, Collart FR, Donnelly MI: A new vector for high-throughput, ligation-independent cloning encoding a tobacco etch virus protease cleavage site. *Protein Expr Purif* 2002, 25:8-15.

45. Donnelly MI, Zhou M, Millard CS, Clancy S, Stols L, Eschenfeldt WH, Collart FR, Joachimiak A: An expression vector tailored for large-scale, high-throughput purification of recombinant proteins. *Protein Expr Purif* 2006, 47:446-454.

46. Dieckman LJ, Zhang W, Rodi DJ, Donnelly MI, Collart FR: Bacterial expression strategies for human angiogenesis proteins. *J Struct Funct Genomics* 2006, 7:23-30.

47. Lin CT, Moore PA, Auberry DL, Landorf EV, Peppler T, Victry KD, Collart FR, Kery V:Automated purification of recombinant proteins: combining high-throughput with high yield. *Protein Expr Purif* 2006, 47:16-24.

48. Deng J, Davies DR, Wisedchaisri G, Wu M, Hol WG, Mehlin C: An improved protocol for rapid freezing of protein samples for long-term storage. *Acta Crystallogr D Biol Crystallogr* 2004, 60:203-204.

49. Niesen FH, Berglund H, Vedadi M: The use of differential scanning fluorimetry to detect ligand interactions that promote protein stability. *Nat Protoc* 2007, 2:2212-2221.

Chapter 3

Seafood Insecurity, Bush Meat Consumption, and Public Health Emergency in West Africa: Did We Miss the Early Warning Signs of an Ebola Epidemic?

Ahmed S Khan[1] and Sanie SS Sesay[2]

[1]School of Business | School of Environment, Saint Mary's University, Halifax, NS, Canada

[2]Department of Clinical Sciences, Liverpool School of Tropical Medicine, Pembroke Place, Liverpool, UK

ABSTRACT

In this article, we frame issues around food security and the Ebola epidemic in West Africa and discuss the potential contribution of interrelated factors like seafood access, bush meat consumption, and public health concerns with the recent outbreak. Since seafood is a major dietary constituent in the affected countries, we posit that seafood unavailability due to unsustainable fishing practices and global change may increase the demand for bush meat and the risk of exposure to zoonosis such as Ebola through hunting and wildlife interactions. We discuss the potential contribution of these contextual drivers to public health within the wider milieu of changing climate, habitat disruptions, human migration and its implication for adaptive capacity, resilience and environmental governance. We conclude with policy options and research directives in addressing regional food security challenges, maritime policy, and emerging global health concerns.

WHAT HAS SEAFOOD GOT TO DO WITH EBOLA?

Since the outbreak of the Ebola epidemic in West Africa almost a year ago, little effort has been directed to understanding the contextual factors and drivers of its emergence and propagation, and how best to address the current crisis and prevent future ones. Most efforts and interventions have been targeted mainly at public health control measures with little attention to the link with food security, regional ecosystem dynamics, and global change. Moreover, this region is comprised of several fragile states and one of the most vulnerable in terms of food security and well-being (Jalloh *et al.* [2013]; Khan and Sei [2015]). Seafood is a major part of the diet in the sub-region (Smith *et al.*[2010]), amounting to 70% of animal protein intake in coastal countries like Sierra Leone (FAO [2014]). The small-scale sector plays a key and important role in this as it strives to address not only food security and well-being but also broader development issues (Neiland and Bene [2004]; Kawarazuka [2010]; Chuenpagdee [2011]). Even with increases in catch landings, the sub-region has been witnessing a net fishery deficit with emphasis on both high-value exports for consumption and low-value for fish meal (Alder *et al.* [2008]; Nunoo *et al.* [2009]). With changing climate and macroeconomic policies that favor exports over local consumption (Kaczynski and Fluharty [2002]; Alder and Sumaila [2004]; Lam *et al.*[2012]), there has being an increase in bush meat hunting since the mid-1970s to the 1990s as shown in Figure 1(Ntiamoa-Baidu [1997]; Brashares *et al.*[2004]; [2011]). This trend has continued as the bush meat trade, estimated in billions of dollars, is a major dietary contributor in West Africa and a lucrative livelihood activity in many parts of the world (Bennett *et al.* [2007]; Golden *et al.* [2011]). The term bush meat (or wild game) is used in contrasts to local animal husbandry or livestocks. Thus in many coastal countries in West Africa, decreasing fish stocks indicate increasing demand for bush meat as an alternative source of animal protein and thus increased potential exposure to zoonosis like Ebola that are found in tropical rain forest ecosystems. Indeed, if fruit bats are potential reservoirs of Ebola and other mammalian species are end hosts (Leroy *et al.*[2005]; Groseth *et al.*[2007]; Barrette *et al.*[2009]), then monitoring bush meat trade and wildlife interactions could potentially reduce the risk of contagion. Conversely, sustainable fishing practices that pay attention to local food security needs will reduce the risk of contact to zoonosis (i.e. hotspots for infectious disease) such as Ebola virus.

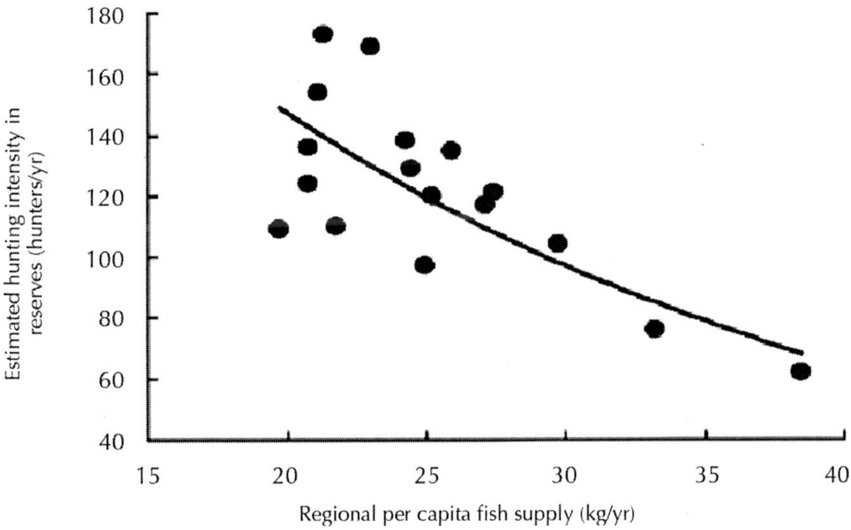

Figure 1: The relationship between per capita fish supply and bush meat consumption (Source: Brashares *et al.*[2004]).

The current epidemic is the worst in human history, about 12, 386 laboratory confirmed cases and about 7,573 deaths as of December 24th 2014[i]. This is expected to increase in the absence of better intervention strategies (WHO Ebola Response Team [2014]). The reaction to this epidemic has been a situation of too little too late, with some medical resources and international commitment only beginning to ensue after the epidemic got out of control (Farrar and Piot [2014]). This has resulted in a huge repercussion on all fabrics of social life: from lock downs to closures of public services such as schools and banks, to broader human development implications including food price surges, regional trade, and the global economy.

[i]CDC. 2014. Case counts. http://www.cdc.gov/vhf/ebola/outbreaks/2014-west-africa/index.html.

Although the international risk of contagion is low in the short term, there are ramifications if there are outbreaks in countries with direct flight connections to West Africa and Europe (Gomes *et al.* [2014]). Hence countries that could potentially be affected are instituting travel restrictions and developing national public health action plans to deal with possible cases. In a concerted effort, humanitarian agencies and ministries of health in the affected countries are soliciting relief and medical supplies to halt the spread of the disease. Are these measures enough to curtail the current epidemic? So far, they don't seem to be working with the cases of Ebola escalating in the region and incidences in other non-neighboring countries. To understand how the current epidemic arose, we need to appreciate the multiple precipitating drivers and recognise that the epidemic is a symptom of much bigger problems stemming from sustainable development needs.

In this article, we first provide an overview of the disease, and its linkages with food security and wildlife management. Next, we argue for a holistic approach in curtailing the spread and preventing future epidemics recognizing the multiple pathways and inter-related factors starting with food and nutritional well-being, wildlife management, climate related changes, and crisis management. Finally, we delve into various governing options and conclude with areas for further research recognizing that protein dietary needs and environmental health are two of several determinants of human well-being.

GENESIS OF THE CURRENT EPIDEMIC IN WEST AFRICA

The Ebola virus can be traced to the Congo as far back as the mid-1970s when it first occurred in a river basin by the same name (WHO [1978]; Groseth *et al.* [2007]; Barrette *et al.* [2009]). Prior to this epidemic in West Africa, previous outbreaks have been reported in countries such as the Congo, Gabon, Russia, Uganda, Italy, US, and the Philippines[ii]. Various species of fruit bat present in sub-Saharan Africa are thought to be reservoirs (Leroy *et al.* [2005]), with apes and swine thought to be end hosts (Barrette *et al.* [2009]; Groseth *et al.*[2007]). Transmission to humans is thought to occur through contact with infected animal tissues and through bodily fluids (WHO Ebola Response Team [2014]). Moreover, there is no proven or approved treatment, although drugs like ZMapp® and other experimental antiretroviral medications have all been used without robust clinical trial. This uncertainty frightens many public health authorities and poses communication challenges to local communities grappling with the epidemic. Like with many other viral haemorrhage fevers (VHFs), there are proven public health control measures like prompt diagnosis and isolation of cases, quarantine of potential cases, surveillance of high risk contacts, safe burial of deceased patients, social distancing and travel restrictions and screening (Feldmann and Geisbert [2011]; WHO Ebola Response Team [2014]).

[ii]CDC. 2014. Outbreak Chronology: Ebola Virus Disease. Accessed Sept. 30th 2014:http://www.cdc.gov/vhf/ebola/outbreaks/history/chronology.html.

There are a number of factors responsible for this epidemic and its rate of progression. First, these countries have not dealt with Ebola before and this led to a lot of initial confusion about the nature of the virus and its transmission. Without the benefit of accurate and prompt diagnosis, it was difficult to differentiate Ebola from other more common locally endemic conditions like malaria and Lassa fever. Second, the international community was slow in recognizing that the epidemic was "a public health emergency of international concern" and to respond and augment adequate resources in the affected countries (WHO Ebola Response Team [2014]). Third, the lack of proven curative treatment, with only experimental therapy and untested vaccines that are controversial thwarts immediate intervention. Furthermore, past incidences of side effects and deaths from trial drugs in the sub-region have led to public mistrust and legal claims (Jegede [2007]; Lenzer[2011]). Fourth, due to the geography of the region, with many populous towns in the Mano River basin bordering Guinea, Liberia and Sierra Leone, cross-border travel increases the risk of contagion from rural to urban areas in the shortest possible time. Finally, this epidemic is a symptom of a much bigger problem: fragile states including Sierra Leone and Liberia with ailing healthcare systems and inadequate health infrastructure. For instance, the doctor to patio ratio in 2010 measured as a density of physicians per 1,000 people is relatively lower in Sierra Leone (0.022) compared to Nigeria (0.408), a non-fragile state[iii].

[iii]Doctor to Patient Ratio in Africa in 2006. Accessed Oct 20th 2014:http://apps.who.int/gho/data/node.main.A1444.

Could we have predicted such epidemics before they occur? We posit that yes, the warning signs were evident. Escalating food prices and declining trend in local food production and low household incomes point to a vulnerable situation (Jalloh *et al.* [2013]). Knowing that Ebola is not the only VHF in the affected region sends heightened signals for public health preparation and action plans (Schoepp *et al.*[2014]). Just two years ago, there was a large epidemic of Cholera in the same region with cross-border spread hinting at the fragile nature of the public healthcare systems. A successful approach to such epidemics should include a synergistic approach between nutritional well-being, public health and environmental governance. Thus, the need to understand the complex and interlinked relationships between seafood and wildlife interactions prompts us to explore response strategies that go beyond sectoral health policies.

INTERPLAY BETWEEN SEAFOOD SECURITY, WILDLIFE MANAGEMENT AND PUBLIC HEALTH

The prevalence of hunger and malnutrition in Sub Saharan Africa has been ranked as "serious" and "alarming", and has implications for infant mortality rate and life expectancy (wheeler and von Braun

[2013]). Although population increase is often seen as a concern for food security (Jalloh *et al.* [2013]), climate change is increasing becoming another threat to reckoned with. Changing climate in the past few decades have consequences on both terrestrial and seafood production systems (Lam *et al.* [2012]; Katikiro and Macusi [2012]; Jalloh *et al.* [2013]). Climate variability has consequences for seafood availability and large marine ecosystem dynamics bordering the affected countries (Hardman-Mountford and McGlade [2003]; Lam *et al.*[2012]). It has been demonstrated that food consumption patterns change with the availability and access of seafood for other protein substitutes such as bush meat. For instance, seafood unavailability has been linked to an increase in wildlife hunters as well as bush meat supply in local markets (Brashares *et al.* [2004]). Recognizing that there is a stronger interactivity between the two protein-based food systems, this has implications for public health through potential contagion with wildlife (Bennett[2002]; Rao and McGowan [2002]; WHO [2003]).

Similarly, climate change has potential impacts on the life cycle of pathogens that affects food production systems and human well-being (Daszak *et al.*[2000]; McMichael *et al.*[2006]; Groseth *et al.*[2007]). Climate-related migrations of pathogens through reservoir hosts and vectors due to habitat disruptions, urbanization, and human migrations are emerging public health concerns (Hale*et al.*[2000]; McMichael *et al.*[2003]). In Western Africa where wildlife populations are historically and increasingly being part of local food systems and economic opportunities (Ntiamoa-Baidu[1997]; Brashares *et al.*[2004]), incidence of zoonosis creates high vulnerability to household food security and public health (Kawarazuka [2010]; Wheeler and von Braun [2013]; Schoepp *et al.*[2014]).

What is discouraging is that most research on the impact of global change on local food systems and infectious diseases look at single causal factors without addressing the multiple pathways, complex interactions, and contextual relationships (Patz *et al.* [2000]; Xun *et al.*[2010]). Despite the body of evidence towards climate variability and pest and diseases and food systems (McMichael *et al.* [2006]), predicting the many causal pathways and trends is challenging due to issues around attributing and contributing factors. Unravelling these multiple factors and their interactions with livelihoods, institutional structures, and decision-making processes could contribute to a better understanding of early warnings, response strategies, and post-crisis management (Dudley [2004]; Altizer *et al.*[2013]). Reactive strategies targeted at preventing such epidemics are likely to be inadequate without elucidating on these complex nature-society interactions (Figure 2).

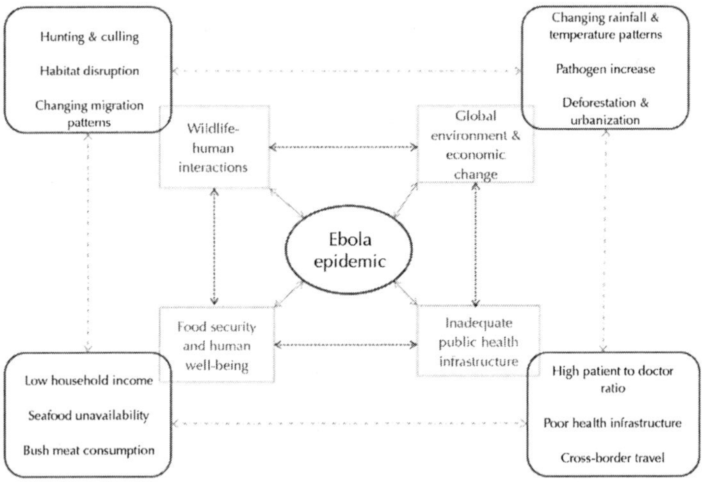

Figure 2: Conceptual linkages in human-nature interactions across marine and terrestrial systems for food security and the implication for Ebola and global public health.

Knowing that food security has been a long term issue in the sub-region, the question becomes how can policy makers improve the adaptive capacity to recover and cope in the short-term as well as being resilient to shocks in the long-term? A healthy community in times of crisis requires an adaptation strategy from being 'vulnerable' to 'resilient' and from 'dis-ease' to 'ease' (Almedom and Tumwine [2008]). It has been suggested that a combination of tools and approaches that draw on food studies, epidemiology, and environment governance is necessary in dealing with multiple entry points and feedbacks (Fraser *et al.*[2003]; McMichael *et al.*[2006], Armitage and Plummer[2010]). Whilst some researchers have argued for a socio-ecological and resilience approach to pandemics (Cumming [2010]), others have underscored nutritional well-being and food secured households (Kawarazuka [2010]; Coulthard [2012]). In other words, "building resilience requires changes in behaviours and norms, transmitted through social [and policy] networks" that fosters local food security

(Frankenberger and Constas [2014]: 744). Given this, making synergies through policy networks especially between environmental health and governance is imperative for social resilience that strengthens well-being (Fraser *et al.*[2003]; Armitage *et al.* [2011]; Broch[2013]).

Xun *et al.* ([2010]) argues that scale presents a bigger challenge in the context of vulnerability assessment (spatial and temporal), as well as bio-complexity in terms of disease ecology and the dynamics in attaining sustainable development. These social interactions across scales and amongst multiple actors are better framed under the rubric of governance. Governance deals with both state and non-state institutions, rules of decision-making, in addition to scale matching (Armitage and Plummer [2010]). It is a mechanism to address food security under uncertainties as well as ecosystem shifts. This is in recognition of the multiple drivers, pathways, and feedback loops necessary for adequately addressing the issues around food production and global change (Folke *et al.*[2005]; Miller *et al.*[2010]). This deserves more focus to which we now turn.

SYNERGIES BETWEEN LOCAL FOOD SYSTEMS AND ENVIRONMENTAL HEALTH

Arguably, vulnerable populations that are food insecure and with an inadequate health care infrastructure signal potential crisis in the event of an epidemic such as EVD. Because EVD is a multi-faceted issue with several entry points, governance mechanisms are necessary for making linkages and synergies across sectors, jurisdictional scales, and interest groups. "Because much of the world's seafood comes from regions with weak governance, improved governance is essential to sustain or increase seafood's contribution to food security" (Smith *et al.* [2010]: 785). Adapting to food or climate related disasters also have an institutional dimension in terms of decision support tools that are intricately connected to multi-level governance arrangements (Kooiman *et al.*[2005], Armitage and Plummer [2010]). These arrangements provide opportunity for linking risk reduction measures in public health with capacity building efforts in promoting well-being (Hess *et al.*[2012]).

There is a lot to learn from the history on governance in food related crisis and public health emergencies. In times of disasters, some scholars have argued that windows of opportunities for social change and transformation do emerge for institutional resilience (Scheffer *et al.*[2002]; Folke *et al.*[2005]; Birkmann *et al.*[2008]). Extreme events have presented opportunities to adapt and cope through strengthening local institutions and training to help manage emerging risks. Previous efforts at curtailing the EVD in Uganda involved rigorous health protocols for isolation, surveillance, health care treatment, supported by local authorities with financial assistance by donor agencies (Borchert *et al.*[2011]). The key lesson is to augment local capacity and resources and strengthen institutional infrastructure. In the case of addressing BSE outbreak in Canada, another zoonotic disease, various adaptation strategies were initiated, including post-crisis coping strategies, food production initiatives, and enhanced engagement with stakeholders (Anderson and McLachlan [2012]).

Building local community resilience that focuses on response and recovery as well as national action plans that promote prevention and preparation is necessary (Keim [2008]). Exemplifying uncertainties due to climate change, inadequate knowledge, and other institutional constraints do provide the basis for adaptive management (Walters [1986]; Nichols *et al.*[2011]). Drawing on diverse disciplinary perspectives, curtailing the epidemic could entail both anticipatory risk reduction measures as well as crisis management through synergistic efforts in addressing food security and environmental health. Moreover, crisis management activities can be made complementary to the adaptive management cycle especially in planning, implementing, monitoring and evaluating action plans in resource management (Figure 3). The governing processes in these interventions could strengthen the science-policy divide; build partnerships, and engage stakeholders in collective action on food security and environmental health (Anderson and McLachlan [2012]).

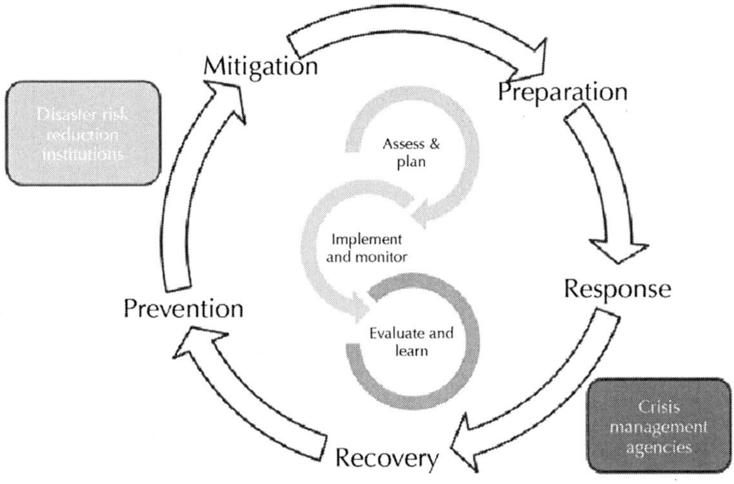

Figure 3: Integrating risk reduction and adaptive measures for food security needs and public health in the case of West Africa. The various colors denote cautionary traffic signals and the arrows denote the need to constantly adjust from disaster planning to post-crisis recovery. Adapted from Folke *et al.* ([2005]); Keim ([2008]); Hess *et al.* ([2012]).

Several models for controlling disease and attaining nutritional well-being have been proposed in the African context (Kock *et al.*[2002]). These include: i) a community-based production system that is local and affordable and focuses on the principle of sufficiency, ii) partnership arrangements between governments and private sector organizations by sharing of benefits and cost and promoting health standards and sanitary research, and iii) a top-down and centralised government agency that promotes and fosters public health policy and the monitoring and surveillance of pathogens.

Other creative governance and flexible decision-making arrangements could be cost-effective for dealing with human well-being through benefit sharing schemes. The Novartis Coartem Anti-malarial Treatment program is an example on a cost recovery drug program spearheaded by industry in partnership with governments, multilaterals, and local partners. This was possible under the Convention on Biological Diversity and Nagoya Protocols on Access and Benefit Sharing arrangements (TEEB [2009]). Also, coordination efforts through regional nodes and risk mapping can be supportive in curtailing the epidemic (Cumming [2010]; Semenza *et al.*[2012]).

Furthermore, a robust regional monitoring scheme is needed. One that entails international travel checks from affected areas due to high mobility of coastal fishers, cross-border land travels, and global airline networks (Tatem [2009]; Binet *et al.*[2012]; Gomes *et al.*[2014]). This could build upon the legacy of the World Health Organization (WHO) for travel protocols such as the Yellow Card and the International Maritime Organization (IMO) on safety and health guidelines[iv].

[iv]IMO Circular Letter No. 3485 10th September - Full and effective implementation of maritime security measures to assist in preventing the spread of EVD.http://www.imo.org/MediaCentre/HotTopics/ebola/Documents/3485.pdf.

We suggest rather than putting emphases on reactive measures such as lock-downs and drug trials as short-term strategies alone, long-term multi-prong approaches that include food security considerations, risk mapping, and environmental monitoring is necessary (Dudley [2004]; Frankenberger and Constas [2014]). Improving bottom-up initiatives for community health and resilience (Berkes and Ross [2013]; Hess *et al.* [2012]) and taking into account broader regional governance and policy directives are crucial (Khan and Mikkola [2002]).

Communicating scientific insights in a region with low literacy in European languages is a task that could benefit from inter-cultural and social learning. Besides, the diversity in beliefs, plethora of world views, and quaint cultural practices make effective communication even more of a priority (Fanon [1961]; Rajnarayan [2002]; Jegede [2007]). Fortunately, this presents an opportunity to bridge science and policy through public engagement and social learning platforms amongst health care professionals, resource managers, policy makers and local community leaders (Kock *et al.*[2002]; Dudley [2004]; Hess *et al.* [2012]; Dietz [2013]).

Social and professional networks are also useful for knowledge mobilization and as bridging organizations for community resilience (Galaz [2009]). This could take the form of institutional partnership and information sharing for public health decision making (Dudley [2004]), remittances for hospital supplies or humanitarian relief, and through public sensitization campaigns (Hess *et al.* [2012]). Volunteer work through the likes of *Doctors without Borders* (MSF) and *Doctors in the Diaspora* could augment local support on the ground. These collective support systems garnered through various networks could be crucial at the local level where resources are limited.

Because the vulnerability of the food systems in the sub-region will affect public health policies and environmental governance, we strongly posit that multiple pathways and entry points present opportunities for various stakeholders to self-organize in addressing the issues either by reducing vulnerability or building resilience.

WAY FORWARD

Coastal fisheries and seafood contributes tremendously to the nutritional well-being of the affected Ebola countries in West Africa. In the immediate absence of vaccines and not knowing whether bush meat and livestocks are tainted (Groseth *et al.* [2007]), seafood is currently providing the necessary caloric intake, essential fatty acids, and micro-nutrients for bolstering the immune systems for those affected to be resilient (Khan and Sei [2015]). Similarly for the healthy, seafood is the main nutritional protein available to be resistant to the disease. The current epidemic is a symptom of complex interplay of social, economic, political, and environmental factors that impact seafood security and changing dietary patterns with consequences on well-being and environmental health. This situation is exacerbated by high poverty levels and regional migration across borders both on land and at sea (Neiland and Bene [2004]; Binet *et al.* [2012]; Gomes *et al.* [2014]). As shown in Figure 2, there are many entry points and pathways in addressing the Ebola issue, including food security and well-being. Understanding how food policies especially sustainable seafood production in meeting local dietary needs benefit public health is central in identifying multiple policy windows for crisis management and future prevention of zoonosis (WHO[2003]; Brashares *et al.* [2004]; Guernier *et al.* [2004]; Barrette *et al.*[2009]).

In the long-term, creating early warning networks, monitoring environmental health, and strengthening local and national food security institutions in promoting adaptive capacity is critical. Attaining collective action through cross-border collaboration, institutional partnerships, and creative governing arrangements at the regional and international level is also important. Knowing the importance of seafood to nutritional well-being and long-term food security in the region, integrating health research with environmental management and regional economic development is also imperative as emerging pathogens are often linked to local and global drivers (Dudley[2004]; Daszak *et al.* [2000]; Barrette *et al.*[2009]). With the International Maritime Organization policy on EVD, vessels exporting seafood from the sub-region are sometimes quarantined for 21 days, thus affecting product quality, fish prices, and regional trade. Equally important is creating institutional synergy and cross-sectoral policies between maritime seafood production and terrestrial wildlife management for avoiding potential environmental risks (Leroy *et al.*[2005]; Richter and Cumming [2006]; Galaz [2009]).

Dietary seafood intake and environmental health are two of many determinants of well-being and community resilience (Kawarazuka [2010]). Recognizing that Ebola is a public health emergency of global proportion, regional working groups and task force on food security and ecosystems spearheaded by UN agencies could provide policy inputs on short and long-term proactive measures. There are many lessons to share with other regions on food security and wildlife interactions, not only for avoiding pandemics, but for innovative institutional mechanisms for addressing future crisis. We may have missed the early warning signs of a catastrophe by ignoring the contribution of food and well-being to public and environmental health, but there is no reason why we cannot curtail this epidemic and prevent future occurrences.

AUTHOR'S CONTRIBUTIONS

Both authors read and approved the final manuscript.

ACKNOWLEDGEMENTS

We are grateful to colleagues who directed us to relevant materials in making connections between seafood, public health and environmental governance especially Professor Firkes Berkes, Dr. Sia Kpakiwa, Dr. Tara Whitty and Dr. John Abraham. Editorial suggestions by Ms. Tiffanie Rainville and Karen Mrema on an earlier manuscript are highly appreciated. We thank Professor Justin Brashares for Fig 1, with reprint permission granted by AAAS. This collaborative piece is dedicated to the memory of Dr. Humar S. Khan aka Squazo, who inspires us to be selfless. Ahmed S. Khan acknowledges financial support during his postdoctoral fellowship from Saint Mary's University, the Social Sciences and Humanities Research Council of Canada and the Natural Sciences and Engineering Research Council of Canada. Sanie S.S. Sesay is supported through the Malaria Capacity Development Consortium funded by the Wellcome Trust.

REFERENCES

1. Alder J, Sumaila UR: Western Africa: a fish basket of Europe past and present. *Journal of Environment and Development* 2004, 13:156-178.
2. Alder J, Campbell B, Karpouzi V, Kaschner K, Pauly D: Forage fish: from ecosystems to markets. *Annual Review of Environment and Resources* 2008, 33:153-166.
3. Almedom AM, Tumwine JK: Resilience to disasters: a paradigm shift from vulnerability to strength. *African Health Sciences (SI on Resilience)* 2008, 8S:1-4.
4. Altizer S, Ostfeld RS, Johnson PTJ, Kutz S, Harvell CD: Climate change and infectious disease: from evidence to a predictive framework. *Science* 2013, 341:514-519.
5. Anderson CR, McLachlan SM: Exiting, enduring and innovating: farm household adaptation to global zoonotic disease. *Global Environmental Change* 2012, 22:82-93.
6. *Adaptive capacity and environmental governance*. Springer, Berlin; 2010.
7. Armitage D, Bene C, Charles AT, Johnson D, Allison EH: The interplay of well-being and resilience in applying a social-ecological perspective. *Ecology and Society* 2011, 17:15.
8. Barrette RW, Metwally SA, Rowland JM, Xu L, Zaki SR, Nichol ST, Rollin PE, Towner JS, Shieh W-J, Batten B, Sealy TK, Carrillo C, Moran KE, Bracht AJ, Mayr GA, Sirios-Cruz M, Catbagan DP, Lautner EA, Ksiazek TG, White WR, McIntosh MT: Discovery of swine as a host for the *Reston ebolavirus*. *Science* 2009, 325:204-206.
9. Bennett EL: Is there a link between wild meat and food security? *Conservation Biology* 2002, 16:590-592.

10. Bennett EL, Blencowe E, Brandon K, Brown D, Burn RW, Cowlishaw G, Davies G, Dublin H, Fa JE, Milner-Gulland EJ, Robinson JG, Rowcliffe JM, Underwood FM, Wilkie DS: Hunting for consensus: reconciling bush meat harvest, conservation, and development policy in West Central africa. *Conservation Biology* 2007, 21:884-887.
11. Berkes F, Ross H: Community resilience: towards an integrated approach. *Society and Natural Resource* 2013, 26:5-20.
12. Binet T, Failler P, Thorpe A: Migration of Senegalese fishers: a case for regional approach to management. *Maritime Studies* 2012, 11:1.
13. Birkmann J, Buckle P, Jaeger J, Pelling M, Setiadi N, Garschagen M, Fernando N, Kropp J:Extreme events and disasters: a window of opportunity for change? Analysis of organizational, institutional and political changes, formal and informal responses after mega-disasters. *Natural Hazards* 2008, 55:637-655
14. Borchert M, Mutyaba I, Van Kerkhove M, Lutwama J, Luwaga H, Bisoborwa G, Turyagaruka J, Pirard P, Ndayimirije N, Roddy P, Van Der Stuyft P: Ebola haemorrhagic fever outbreak in Masindi District, Uganda: outbreak description and lessons learned. *JBMC Infectious Diseases* 2011, 11:357.
15. Brashares JS, Arcese P, Sam MK, Coppolillo PB, Sinclair ARE, Balmford A: Bush meat hunting, wildlife declines, and fish supply in West Africa. *Science* 2004, 306:1180-1183.
16. Brashares JS, Golden CD, Weinbaum KZ, Barrett CB, Okello GV: Economic and geographic drivers of wildlife consumption in rural Africa. *PNAS* 2011, 108:13931-13936.
17. Broch HB: Social resilience - local responses to changes in social and natural environments. *Maritime Studies* 2013, 12:6.
18. *World small-scale fisheries contemporary visions*. Eburon, Delft; 2011.
19. Coulthard S: Can we be both resilient and well, and what choices do people have? Incorporating agency into the resilience debate from a fisheries perspective. *Ecology and Society* 2012, 17(1):4.
20. Cumming GS: Risk mapping for avian influenza: a social–ecological problem. *Ecology and Society* 2010, 15:32.

21. Daszak P, Cunningham AA, Hyatt AD: Emerging infectious diseases of wildlife- threats to biodiversity and human health. *Science* 2000, 287:443-449.

22. Dietz T: Bringing values and deliberations to science communication. *PNAS* 2013, 110:14081-14087.

23. Dudley JP: Global zoonotic disease surveillance: an emerging public health and biosecurity imperative. *BioScience* 2004, 54:982-983.

24. Fanon F: *The wretched of the earth*. Groove Press, New York; 1961.

25. FAO: *The state of world fisheries and aquaculture 2013. Challenges and opportunities*. FAO, Rome; 2014.

26. Farrar, JJ, and P Piot. 2014. The Ebola emergency – immediate action, ongoing strategy. New England Journal of Medicine Doi:10.1056/NEJMe1411471.

27. Feldmann H, Geisbert TW: Ebola haemorrhagic fever. *Lancet* 2011, 377:849-862.

28. Folke C, Hahn T, Olsson P, Norberg J: Adaptive governance of social-ecological systems. *Annual Review of Environment and Resources* 2005, 30:441-473.

29. Frankenberger TR, Constas MA: IFPRI 2020 conference: building resilience for food and nutritional security, Addis Ababa, 15–17 of May 2014: a synthesis. *Food Security* 2014, 6:743-745.

30. Fraser EDG, Mabee W, Slaymaker O: Mutual vulnerability, mutual dependence: the reflexive relation between human society and the environment. *Global Environmental Change* 2003, 13:137-144.

31. Galaz V: Pandemic 2.0: Can information technology help save the planet?*Environment* 2009, 5:20-28.

32. Golden DC, Lia C, Fernald H, Brashares JS, Rasolofoniaina BJR, Kremen C: Benefits of wildlife consumption to child nutrition in a biodiversity hotspot. *PNAS* 2011, 49:19653-19656.

33. Gomes, MFC, A Pastore y Piontti, L Rossi, D Chao, I Longini, ME Halloran, and A Vespignani. 2014. Assessing the international spreading risk associated with the 2014 West African Ebola outbreak. PLOS Currents Outbreaks. Doi:10.1371/currents.outbreaks.cd818f63d40e24aef769dda7df9e0da5

34. Groseth A, Feldmann H, Strong JE: The ecology of Ebola virus. *Trends in Microbiology* 2007, 15(9):408-416.
35. Guernier V, Hochberg ME, Guégan J-F: Ecology drives the worldwide distribution of human diseases. *PLoS Biology* 2004, 2(6):e141.
36. Hale S, Kovats S, Woodward A: What El Nino can't tell us about human health and global climate change? *Global Change Human Health* 2000, 1:66-77.
37. Hardman-Mountford NJ, McGlade JM: Seasonal and inter-annual variability of oceanographic processes in the Gulf of Guinea: an investigation using AVHRR sea surface temperature data. *International Journal of Remote Sensing* 2003, 24(16):3247-3268.
38. Hess JJ, McDowell JZ, Luber G: Integrating climate change adaptation into public health practice: using adaptive management to increase adaptive capacity and build resilience. *Environmental Health Perspectives* 2012, 12o:171-179.
39. *West African agriculture and climate change: a comprehensive analysis*. International Food Policy Research Institute, Washington, DC; 2013.
40. Jegede AS: What led to the Nigerian boycott of the polio vaccination campaign? *PLoS Medicine* 2007, 4(3):e73. doi: 10.1371/journal.pmed.0040073
41. Kaczynski V, Fluharty DL: European policies in West Africa: Who benefits from fisheries agreements? *Marine Policy* 2002, 26:75-93.
42. Katikiro RE, Macusi ED: Impacts of climate change on West African fisheries and its implications on food production. *Journal of Environmental Science and management* 2012, 15:83-95.
43. Kawarazuka N: *The contribution of fish intake, aquaculture, and small-scale fisheries to improving nutrition: a literature review. The WorldFish center working paper No.2106*. The WorldFish Center, Penang, Malaysia; 2010.
44. Keim ME: Building human resilience: the role of public health preparedness and response to as an adaptation to climate change. *American Journal of Preventive Medicine* 2008, 35:508-516.

45. Khan AS, Mikkola H: *Sustainable ocean development: an initiative for the management and protection of the marine and coastal resources of the Economic Community of West Africa States (ECOWAS) Sub region*. University of the Gambia, UNIGAM 0001, Banjul; 2002.
46. Khan AS, Sei S: The co-governance of fisheries in post-conflict Sierra Leone: Is the transition for better or for worse? In *In Interactive Governance for Small-scale Fisheries: Global Reflections, eds. S Jentoft, R Chuenpagdee. MARE Publication Series*. Springer, Amsterdam; 2015. [Forthcoming]
47. *Fish for life: interactive governance for fisheries*. Amsterdam University Press, Amsterdam; 2005.
48. Kock R, Kebkiba B, Heinomen R, Bedane B: Wildlife and pastoral society – shifting paradigms in disease control. *Annals New York Academy of Sciences* 2002, 969:24-33.
49. Lam VWY, Cheung WWL, Swartz W, Sumaila UR: Climate change impacts on fisheries in West Africa: Implications for economic, food and nutritional security. *African Journal of Marine Science* 2012, 34:103-117.
50. Lenzer J: Pfizer settles with victims of Nigerian drug trial. *British Medical Journal* 2011, 343:d5268. doi: 10.1136/bmj.d5268
51. Leroy EM, Kumulungu B, Pourrut X, Rouquet P, Hassanin A, Yaba P, Delicat A, Paweska JT, Gonzalez J-P, Swanepoel R: Fruit bats as reservoirs of Ebola virus. *Nature* 2005, 438:575-576.
52. *Climate change and human health: risks and responses*. WHO, Geneva; 2003.
53. McMichael AJ, Woodruff RE, Hales S: Climate change and human health: present and future risks. *The Lancelet* 2006, 367:859-869.
54. Miller K, Charles A, Barange M, Brander K, Gallucci V, Gasalla MA, Khan AS, Munro G, Murtugudde R, Ommer RE, Perry PI: Climate change, uncertainty, and resilient fisheries: Institutional responses through integrative science. *Progress in Oceanography* 2010, 87(1-4):338-346.
55. *Poverty and small-scale fisheries in West Africa*. Springer, New York; 2004.

56. Nichols JD, Koneff MD, Heglund PJ, Knutson MG, Seamans ME, Morton JM, Jones MT, Boomer GS, Williams BK: Climate change, uncertainty, and natural resource management. *Journal of Wildlife Management* 2011, 75:6-18.

57. Ntiamoa-Baidu Y: *Wildlife and food security in Africa*. FAO Conservation Guide No. 33. FAO, Rome; 1997.

58. Nunoo FKE, Boateng JO, Ahulu AM, Agyekum KA, Sumaila UR: When trash fish is treasure: the case of Ghana in West Africa. *Fisheries Research* 2009, 96:167-172.

59. Patz JA, Graczyk TK, Geller N, Vittor AY: Effects of environment change on emerging parasitic diseases. *International Journal of Parasitology* 2000, 30:1395-1405.

60. Rajnayaran C: Plague panic and epidemic politics in India, 1896–1914. Epidemics and Ideas: Essays on the Historical Perception of Pestilence (1992). In *The origins of industrial capitalism in India: business strategies and the working classes in Bombay, 1900–1940*. Cambridge University Press, Cambridge; 2002.

61. Rao M, McGowan PJK: Wild meat use, food security, livelihoods, and conservation. *Conservation Biology* 2002, 16:580-583.

62. Richter HV, Cumming GS: Food availability and the annual migration of the straw coloured fruit bat (Eidolon helvum) at Kasanka National Park, Zambia. *Journal of Zoology, London* 2006, 268:35-44.

63. Scheffer M, Westley F, Brock WA, Holmgren M: Dynamic interaction of societies and ecosystems – linking theories from ecology, economy and sociology. In *Panarchy: understanding transformations in human and natural systems*. Edited by Gunderson LH, Holling CS. Island Press, London; 2002:195-240.

64. Schoepp RJ, Rossi CA, Khan SH, Goba A, Fair JN: Undiagnosed acute viral febrile illnesses, Sierra Leone. *Emerging Infectious Disease* 2014, 20:1176-1182.

65. Semenza JC, Suk JE, Estevez V, Ebi KL, Lindgren E: Mapping climate change vulnerabilities to infectious diseases in Europe. *Environmental Health Perspective* 2012, 120:385-392.

66. Smith MD, Roheim CA, Crowder LB, Halpern BS, Turnipseed M, Anderson JL, Acshe F, Bourillon L, Guttormsen AG, Khan AS, Liguori LA, McNevin A, O'Connor MI, Squires D, Tyedmers

P, Brownstein C, Carden K, Klinger DH, Sagarin R, Selkoe K: Sustainability and global seafood. *Science* 2010, 327:784-786.

67. Tatem AJ: The worldwide airline network and the dispersal of exotic species: 2007–2010. *Ecography* 2009, 32:94-102.
68. TEEB: *2009. The economics of ecosystems and biodiversity for business report*. UNEP, Geneva and Nairobi; 2009.
69. Walters CJ: *Adaptive Management of Renewable Resources*. McGraw Hill, New York; 1986.
70. Wheeler T, von Braun J: Climate change impacts on global food security. *Science* 2013, 341:508-513.
71. WHO: Ebola hemorrhagic fever in Zaire, 1976. Report of an International Convention *Bulletin of the World Health Organization* 1978, 56(2):271-293.
72. WHO: *Climate change and human health: risk and responses*. WHO in collaboration with UNEP and WMO, Geneva; 2003.
73. WHO Ebola Response Team. 2014. Ebola Virus Disease in West Africa. The first 9 months of the epidemic and forward projections. New England Journal of Medicine Doi:10.1056/NEJMoa1411100
74. Xun WW, Khan AE, Michael E, Vineis P: Climate change epidemiology: methodological challenges. *International Journal of Public Health* 2010, 55:85-96.

Chapter 4

Scaling of Wireless Sensor Network Intrusion Detection Probability: 3D Sensors, 3D Intruders, and 3D Environments

Omar Said[1, 3] and Alaa Elnashar[2, 3]

[1]Department of Computer Science, College of Science, Menofia University, Shebin El Kom, Gamal Abdelnaser St., Menofia 32512, Egypt

[2]Department of Computer Science, College of Science, Minia University, Ibrahimia St., Minia 61519, Egypt

[3]College of Computers and Information Technology, Taif University, Al-Hawyia 21974, Taif, Saudi Arabia

ABSTRACT

In this paper, a new model that deploys heterogeneous sensors in 3D wireless sensor networks (WSNs) is proposed. The model handles the two sensing scenarios, single sensing and multiple sensing. The probabilities of intrusion detection in a 3D environment with sensors distributed using Gaussian, uniform, beta, and chi-square are compared. The resultant probabilities values help the WSN security designers in selecting the most suitable sensors deployment regarding some critical network parameters such as quality-of-service (QoS). WSN efficiency under different probabilistic distributions is also demonstrated. To evaluate the proposed model, a simulation environment is constructed using OPNET and NS2. The simulation results showed that Gaussian distribution provides the best efficiency and performance.

INTRODUCTION

Wireless sensor networks (WSNs) consist of small main components which have limited power devices such as sensors and can be installed in open environments [1]-[3]. So, WSNs may be attacked by intruders. Since WSNs applications are used in different fields such as environmental sensing, industrial monitoring, and military process management, intrusion detection became an extremely important issue [4],[5]. In WSN, a huge number of sensors should be deployed for intruder detection. However, the high cost of this solution makes it impractical. Furthermore, using a huge number of sensors does not guarantee a successful detection of a moving intruder within a certain distance since void area may be found in the WSN. There are two main categories for intrusion detection problem. The first one uses a component to monitor WSN security. This component may be software, hardware, or human. The target of this component is accomplished by using some sensors to ensure that the security level in WSN is acceptable [6]-[8]. The second one detects the intruder when it tries to storm unauthorized area [9]-[13]. The time consumed for intruder detection process is an important parameter that should be considered. Accordingly, the intruder should be detected at the same time of its entrance. So, raising the probability of intruder detection in WSNs is concerned to sensor deployment plan more than the number of sensors. Also, wired network contains many intrusion detection systems which are not accommodating the nature of WSNs [14]-[17]. Consequently, developing an innovative technique that deals with WSNs nature is an important issue. Furthermore, finding the optimal representation of sensor deployment that provides the best intruder detection is another important issue. In this paper, a model for intrusion detection in 3D environments is introduced. The model uses various probability distributions to deploy sensors within the entire WSN. A simulator is created to simulate the intrusion detection process and evaluate its efficiency based on the probability distribution used. This will guide the WSN designer to select the optimal sensors distribution that yields the best intrusion detection efficiency. The paper is organized as follows: in Section 3, the problem definition is introduced. In Section 3, the related works are demonstrated. In Section 3, the proposed model is presented. In Section 3, the simulation environment is constructed and the results are discussed. Finally, conclusion and future work are introduced in Sections 3 and 7, respectively.

PROBLEM DEFINITION

WSNs consist of large number of inter communicated sensors via wireless interfaces. These sensors may have various processing capabilities, computing resources, or coverage scenarios. Many WSN applications contain different types of sensors which are used in different types of tasks. Also, there are many applications that are not only used in 2D environment but also in 3D space. The problem of intrusion detection has not been studied extensively in the case of heterogeneous WSNs sensors and also in the case of 3D environments because of its complexity. Also, the effect of mobile sensors on the heterogeneous WSNs is not studied. So, till now, there is no standard model that helps security designers in selecting a suitable sensors distribution that accommodates the nature of WSN 3D applications and available network specifications.

The problem is defined as follows: find the best distribution of network sensors that minimizes the probability value $\Pr(V_s \subset (U^n_{m-1} V_{Im}))$ or maximizes $\Pr((U^n_{m-1} V_{Im}) \cap V_s = \phi$, where *VS* is the space volume that is covered by a network sensor S, and *VI* is the space volume that is covered by an intruder I, see Figure 1.

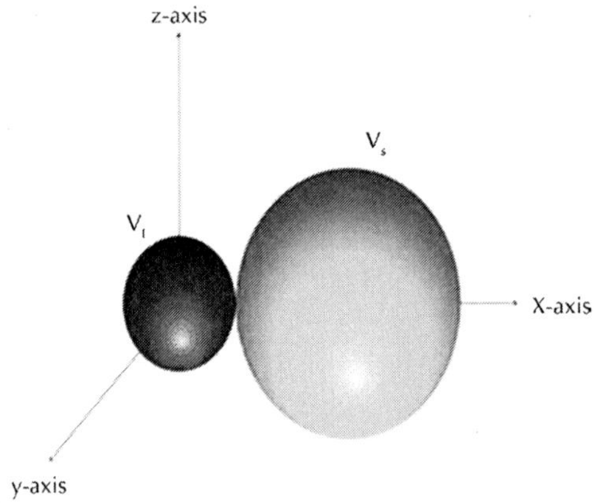

Figure 1: Sensor/intruder spaces in 3D.

For simplification, suppose one sensor *S* and one intruder *I*. The sensor *S* covers the volume *VS*that is represented by a sphere of radius *RS* centered at *VS* center=(x_1,y_1,z_1). The intruder *I*covers the volume *VI* that is represented by a sphere of radius *RI* centered at *VI* center=(x_2,y_2,z_2).

RELATED WORKS

WSN intrusion detection is a challenging process since it can be affected by several parameters such as number of sensors, sensor types, quality-of-service (QoS), detection cost, and WSN environment. Few studies have been introduced to solve the intrusion detection problem in different environments with different types of sensors. These studies can be classified into two categories. The first one manipulates the detection problem with different factors rather than the factors that are presented in this paper. The second category is close related work, which tried to study this problem under nearby parameters of the proposed idea but with different problem definitions and models.

General Related Work

Several models and studies were presented to handle intrusion detection problem. Anomaly-detection-based model with statistical analysis [18] was introduced for intrusion detection in *ad hoc*network. The model takes a long time to detect the intruder due to huge transmitted data and traffic. In [19], a framework, which enables the network designer to select the optimized intruder detection system according to his particular needs, is proposed. In [20], a method for detecting packet modification attacks and supporting target node location privacy is proposed. Moreover, accuracy of this proposed method is evaluated using small-scale WSN. Detection of a moving intruder in a curved path was studied in [21]. The technique uses sine function to define the moving curve of the intruder with detection probability in WSN. Regarding heterogeneous WSNs with the Gaussian distribution, it was proved that the probability of intrusion detection (PID) increases as the number of sensors increases [22]. The probability of intruder detection in homogeneous and heterogeneous WSNs with consideration of density and sensing range for each WSN node was studied in [23]. A comparison between the performance of WSN with Gaussian and Poisson probability distribution under different network settings was presented in [24]. The area covered by WSN at the exact time or within certain time interval in addition to calculating the time that have been consumed to detect a randomly sited inactive object was proposed in [25]. In [26], the relation between the intrusion detection time and the distance, which the intruder cuts down from the start point until hacking the field of interest, is determined. Furthermore, [27] showed that lack of sensors can be compensated by sensor mobility which improved network coverage. Some other ideas such as furtive wireless communication have been demonstrated in [28].

Closely Related Work

There are three close researches to the proposed idea. The first one [29] implemented Gaussian and uniform distributions to determine the probability of intruder detection in WSN. The proposed technique in distinguished between the detection probability as regards the requirements of each WSN application and the network parameters. In addition, the proposed system was studied under two scenarios: single sensing detection and multiple-sensing detection. But this proposed technique did not study the 3D environment under heterogeneous WSN. The second close research proposed a technique that combines Gaussian and Poisson probability distributions [30]. Gaussian distribution was applied to the central area covered by WSN. Poisson distribution was applied to the remaining area. Therefore, this technique is considered as a mixture of two probability distributions in one test area. Also, this technique supposed that the WSN contains heterogeneous sensors. But the tested environment was not in 3D space. Furthermore, it should test each probability distribution independently on the WSN area before proposing a mixture of them. The last one proposed an analysis for intrusion detection problem in a 3D environment that is represented by a cube with two types of sensors [31]. This technique has some drawbacks such as the probability density function is not calculated and a cube is not sufficient to prove the model results. In addition, this proposed model is not well defined.

THE PROPOSED MODEL

Since many applications are applied in 3D environment, optimizing sensor distribution to enhance intrusion detection probability should be considered. In addition, sensor heterogeneity is also another important parameter that should be well studied since most of the intrusion detection researches focus on using homogenous sensors. The main purpose of the proposed model is to analyze the intrusion detection probability, which helps in sensor deploying, to gain a satisfied WSN security level. The effect of the used probability distribution such as Gaussian, uniform, beta, and chi-square are also studied. Figure 2a,b shows samples from sensors distribution views using uniform and normal. Reducing communication redundancy can be achieved by intelligent sensor deployment in WSN. The proposed model deals with four components, a 3D environment within which the WSN should be installed, 3D sensors that are used to detect an intruder, 3D intruder(s) which is (are) predicted to hack WSN, and a target resource that is assumed to be protected by WSN sensors. The 3D environment is represented by a cubic space with predefined dimensions. Sensors, intruders, and targets are presented as sphere centers as shown in Figure 3a. Sensors are heterogonous since each sensor is assumed to be located at a sphere center with a predefined radius that may differ from other sensors radii as shown in Figure 3b. Each sensor covers its entire sphere volume space and is allocated dynamically. Each intruder is also represented as in case of sensors, and its sphere space represents the intrusion space as shown in Figure 3c. Target space is the sphere volume that is required to be protected. The target is hacked by a specific intruder if it is located outside any of sensors coverage spaces, and its distance from the intruder is less than or equal the sum of the intruder sphere and its sphere radii as shown in Figure 3d; otherwise, it is considered to be safe as shown in Figure 3e,f. The steps of the proposed model are described in Algorithm 1 shown below.

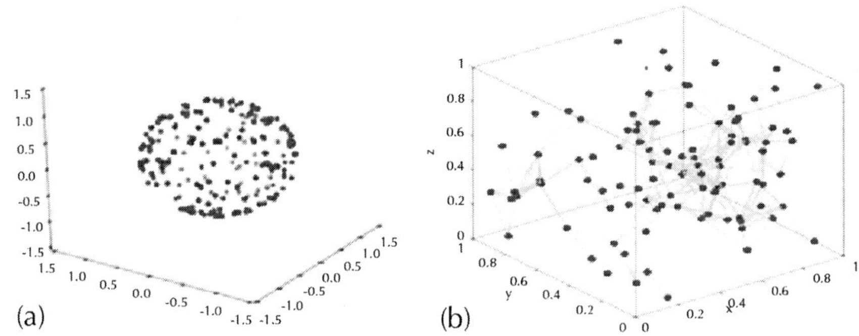

Figure 2: Samples from sensor distribution views.(a) Sensors with uniform distribution in 3D [32]. (b) Sensors with normal distribution in 3D [33].

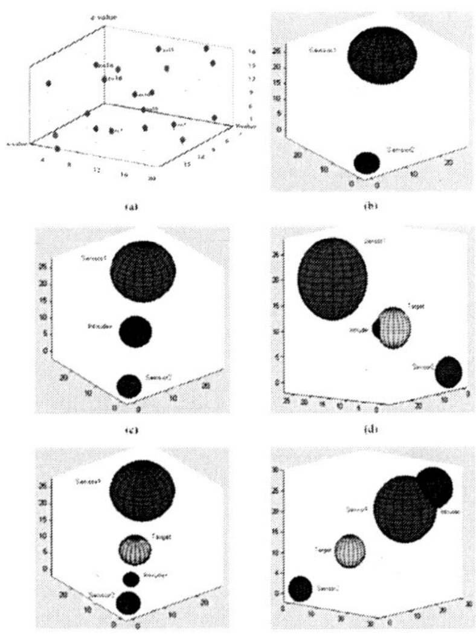

Figure 3: 3D environment with sensor distributions and samples of intruder/sensors/target states. (a) 3D environment with sensors spheres centers. (b)

Two heterogeneous sensors coverage spaces. (c)Two heterogeneous sensors and an intruder representation (d) Hacked target. (e) Safe target - no intruders detected. (f) Safe target - an intruder detected.

SIMULATION AND EVALUATION

A simulator has been developed to simulate the intrusion detection process with multiple intruders in 3D environment with sensors that are distributed by using various probability distributions. The simulator is also designed to evaluate the proposed model. The evaluation is based on the effect of the used probability distributions on efficiency of WSN and also the network parameters such as packet loss, end-to-end delay, and throughput. The simulation parameters and their values are listed in Table 1.

Table 1: Simulation parameters and their values

Simulation parameter	Value
Packet size	1,024 Bytes
Destination	Random
Packet inter-arrival time	Constant (1.0)
Simulation time	2 h
Simulation time	2 h
Battery life	3 to 4 h [active sensing]
	and 1,000+ h [standby]
Start time	Uniform (20, 21)
Transmission range	1 to 100 m
ACK mechanism	Enabled
3D camera	170 (W) × 54 (H) × 49 (D) mm
MAC layer	802.15.4
Transmit power	0.05 mw/m ^{3}
Transmit band	2.4 GHz
Environment space volume	10 × 10 × 10 km ^{3}

Said and Elnashar

Said and Elnashar *EURASIP Journal on Wireless Communications and Networking* 2015 2015:46, doi: 10.1186/s13638-015-0240-6

OPNET 14.5 and NS2 [34], [35] are used to simulate a 3D environment that contains ZigBee with its three layers: application, media access control (MAC), and network. In the MAC layer, the 802.15.4 MAC protocol is used in addition to the ZigBee (CSMA/CA) model [34]. The network layer is implemented by the ZigBee network model, which is used in the routing and request handling.

In the simulated environment, each 3D camera is attached to one sensor to reflect the 3D nature. The simulator is built on large scale as shown in Table 1 and includes different numbers of Zigbee end devices, coordinators, and routers. WSN sensors are communicated via infrared media using star topology and distributed using Gaussian, uniform, beta, and chi-square. Configuring WSN model components and selecting the scalable parameters for the entire simulated WSN are accomplished.

Intruder Detection Efficiency Evaluation

Suppose that the volume space covered by a network sensor is *Vs* and the volume space covered by intruder is *Vi*. $Pr(Vi \cap Vs)$ is the coverage spaces intersection probability of an intruder and a sensor. If this probability equals ϕ, then the system is safe and there is a distance between the intruder and its target. So, this probability should be maximized. $Pr(Vi \cap Vs)$ is the worst probability because the intruder is around to reach its target and hack the security system. So, this probability should be minimized. The distance between the target and an intruder and also its velocity are two factors that are used to determine the time consumed by the intruder to hack the target. Knowing this time value enables the security system manager to protect the target against hacking. In the proposed simulation, the values of intruders' velocities and their distances from the target are randomly generated. Equations 11 to 4 are for Gaussian, uniform, beta, and chi-square probability distributions, respectively [29],[36],[37]. Equation 5 describes how to extract the time from the intruder velocity and the current distance between the intruder and his target [38]. Equation 6 describes the relationship between the intruder and the sensors spheres in the security domain[39].

To evaluate the efficiency of intruder detection process, a simulator with 800 3D sensors covering 3D environment and two 3D intruders is used. The positions of both the target and intruders are created randomly in the simulated environment. The distance between the intruder and the target in addition to the distance between the intruder and the nearest sensor to the target are considered [26].

Two experiments were carried out using this simulated environment. In the first experiment, the values of the distance between the intruder and the target and also the distance between the intruder and the nearest sensor to the target are ignored, in contrast to the second one that considered these values.

Figure 4 shows the simulation results of the first experiment. The results show that Gaussian sensor distribution provides the best average performance for all the numbers of sensors except for 400 sensors. A lower performance is provided in case of using uniform sensor distribution especially for 100 and 700 sensors. Beta and chi-square distributions have the lowest average performance values. Figure 5 shows the simulation results of experiment 2. The results show that Gaussian sensors distribution provides the best average performance for all the numbers of sensors. A lower performance is provided in the case of using uniform sensor distribution. Beta and chi-square distributions have the lowest average performance values.

Algorithm 1

```
Dm is the minimum distance
I: Intruder and T: Target
M: The number of intruders
N: The number of sensors
DjT: Intruder/Target distance
DTSi: Nearest Sensor/Target distance
Suppose Dm = 0, F = 0.
I position is random.
T position is random.
Let Dm = X
The nearest sensor from target is in the same direction of
intended intruder.
For each Gaussian, uniform, beta, and chi-square
Begin
While (F ≺ M)
Begin
For j = 1 TO M
Begin
For i = 1 to N
          Begin
          If (Dm ≻ DIT )
             Dm = DIT
          End
Si = 1 (Flaged)
If (DjT = = DTSi)
Print 'Middle State.'
ElseIF DjT ≻ DTSi
Print 'Save State.'
ElseIF DjT ≺ DTSi
Print 'Hacking State.'
End
For K = 1 to M
Begin
If ((Ik ∩ Sj ≠ Φ ), j=2 to n)
          Print 'Detection process is distributed over multi
sensors
          depending on the overlapped volume spaces.'
End
F = F+1.
End
End
```

Figure 4: The simulation results of the first experiment. Number of successful intruders for each distribution without considering distance and velocity parameters.

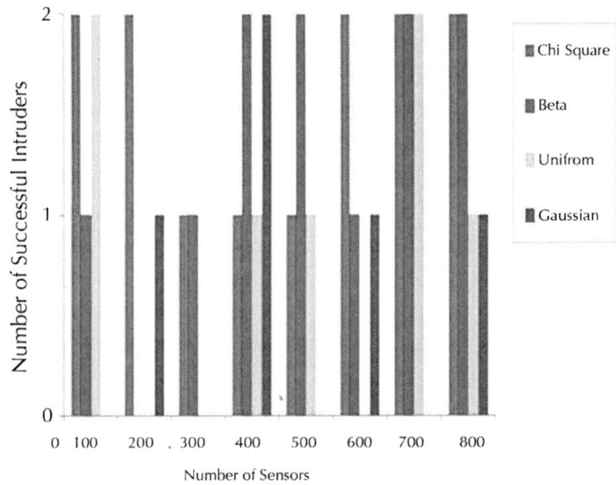

Figure 4: The simulation results of the second experiment. Number of successful intruders for each distribution with considering distance and velocity parameters.

Gaussian [29]

$$f(x,\mu,\sigma) = \frac{1}{\sigma\sqrt{2\Pi}} e^{\frac{(x-\mu)^2}{2\sigma^2}} \tag{1}$$

Uniform [29]

$$f(x,k) = \begin{cases} \frac{1}{u-1} & if \in R_x \\ 0 & if \notin R_x \end{cases} \tag{2}$$

Beta [36]

$$f(x,\alpha,\beta) = \frac{1}{B(\alpha,\beta)} x^{\alpha-1}(1-x)^{\beta-1} \tag{3}$$

Chi-square [37]

$$f(x,k) = \begin{cases} \frac{x^{\frac{k}{2}-1} e^{\frac{-\Pi}{2}}}{2^{\frac{k}{2}}\Gamma\left(\frac{k}{2}\right)} & \\ 0 & x \geq 0 \end{cases} \tag{4}$$

Time extraction [38]

$$T = \frac{D}{V} \tag{5}$$

Spheres intersection [39]

$$\frac{4d^2R^2 - \left(d^2 - r^2 - R^2\right)^2}{4d^2} \tag{6}$$

Evaluation of WSN Sensor Distribution

It is obvious that if the WSN efficiency is not acceptable, the intruder may be detected but the security system may fail to recover this problem. To complete a test of the proposed model efficiency, the cycle after the intruder detection process should be evaluated. This cycle means that data, which are sent to inform the security manager with intruder, should be transmitted successfully. So, the WSN efficiency should be evaluated. The evaluation parameters are the average number of control bits, the number of hops, the average number of lost packets, the end-to-end delay, the throughput, and the general efficiency. In the following subsections, the simulation environment is constructed and each parameter results is showed and discussed.

Average Number of Control Bits

During simulation execution, trace files containing the number of control bits are created instantly for the number of used sensors. The control bits are the retransmitted bits, the ACK bits and the bits which are used to initiate the sessions between WSN sensors in case of sudden event occurrence such as intruder detection. In general, as the number of management bits decreases, the energy consumption decreases and sensors battery live increases.

Figure 6 shows the transmission control data for different WSNs sensors distributed using Gaussian, uniform, beta, and chi-square. The simulation results indicate that Gaussian WSN has the minimum number of management bits compared with other WSNs in contrast to chi-square WSNs that have the maximum number of control bits.

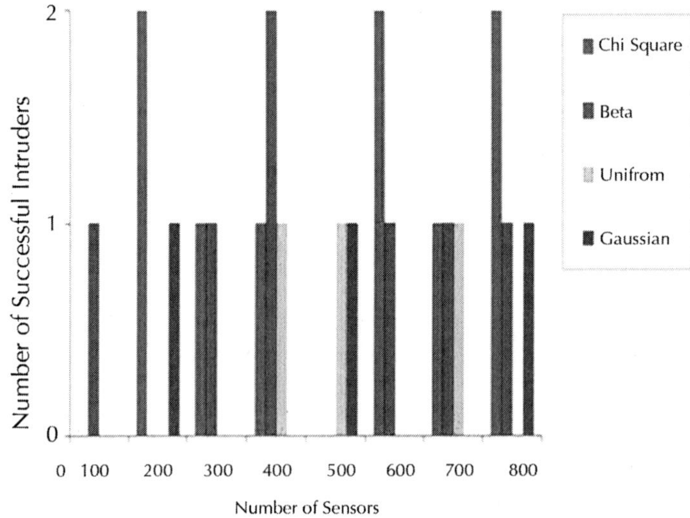

Figure 6: Transmission control data for different WSNs (Gaussian, uniform, beta, and chi-square).

The Number of Hops

The routing path is determined by the number of hops, which are used in data transmission. So, the number of transmission hops is an important factor in determining the best routing paths and their alternatives.

Figure 7 shows that the number of hops used in the Gaussian WSN is less than the number of hops used in the other WSNs (uniform, beta, and chi-square). When the number of sensors equals 600, the Gaussian WSN and the uniform WSN have the same number of transmission hops. The hesitations in curve plots are owed to the randomly chosen sources and destinations in the WSNs. In addition, failure of multiple sensor nodes may cause data collision which requires alternative nodes to complete the transmission process that increases the number of hops. Furthermore, when the number of sensors is greater than 300, the plots' values are decreased. This occurs because the data, which are transmitted within the WSN at this number of sensors, use the best routing path. The simulation results indicate that the Gaussian WSN uses the minimum number of hops compared with the other WSNs. In contrast, the beta and the chi-square WSNs have the maximum number of used hops.

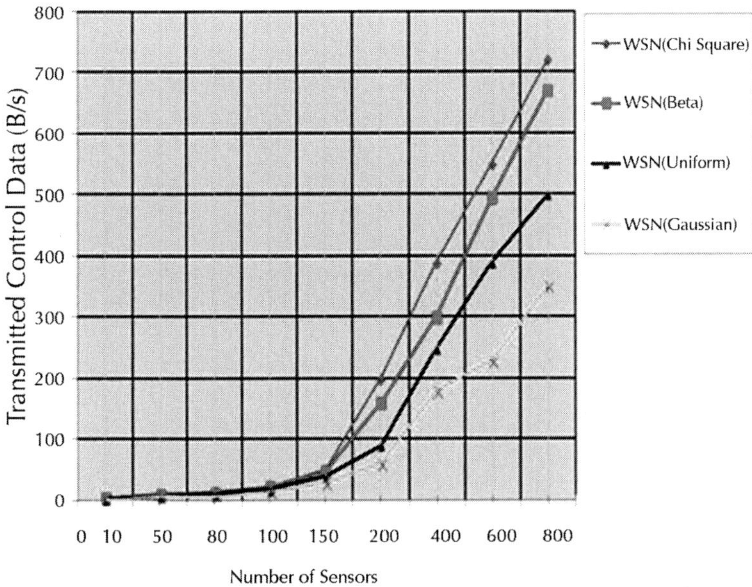

Figure 7: Average number of hops for different WSNs (Gaussian, uniform, beta, and chi-square).

The Average Number of Lost Packets

In the general networks, it is well known that the transmitted data increases when the number of nodes increases which may cause more packet loss [40]. In WSNs, there is a slight difference; the amount of transmitted data depends on the sensor-acquiring information when urgent events occurred [40]. Also, the number of lost packet affects the network efficiency. So, the number of packet loss should be monitored. Figure 8 shows the average number of lost packets for different WSNs sensors distributed by using Gaussian, uniform, beta, and chi-square distributions. The simulation results indicate that the number of sensors does not negatively affect the number of lost packets for a medium number of sensor (≤300) for all simulated WSNs. For a large number of sensors (>300), the average number of lost packets exponentially increases as the number of sensors increases in case of beta and chi-square WSNs in contrast to Gaussian and uniform WSNs, in which the average number of lost packets increases slowly as the number of sensors increases.

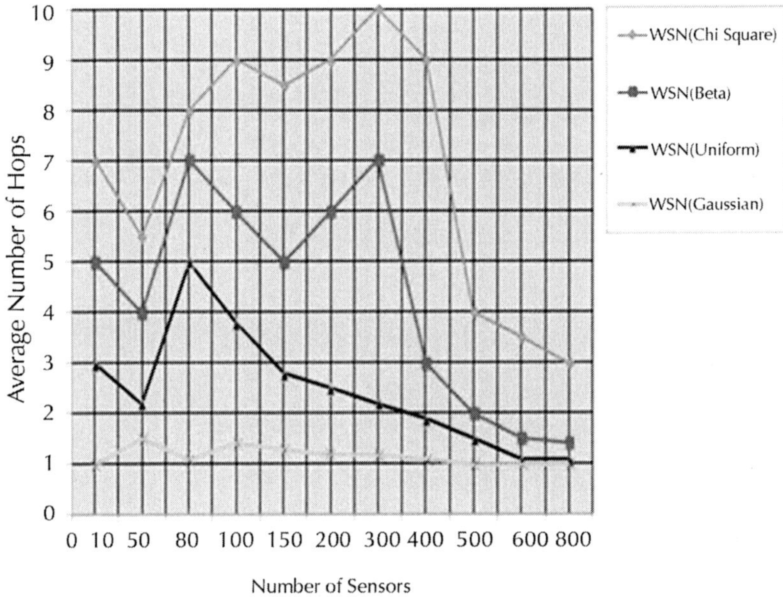

Figure 8: Average number of packet loss for different WSNs (Gaussian, uniform, beta, and chi-square).

Average End- to-end Delay

The end-to-end delay is defined as the total time, which is consumed in packet transmission from source to destination [40]. The end-to-end delay is measured by a time difference between message buffering time at the source and the time of receiving the last bit at the destination. Figure 9 shows the average end-to-end delay for different WSNs sensors distributed by using Gaussian, uniform, beta, and chi-square distributions. The simulation results indicate that uniform and Gaussian WSNs have the minimum average delay, respectively, compared with the other WSNs. Chi-square and beta WSNs have the maximum delay. The interpretation for these results is that in case of the uniform and the Gaussian WSNs, little number of hops are used to transmit the data from source to destinations; this decreases the total time of hops data handling which leads to less end-to-end delay. On the other hand, the beta and the chi-square WSNs routing paths use more hops to transmit the required data from source to destinations (including sinks) which leads to an increased end-to-end delay.

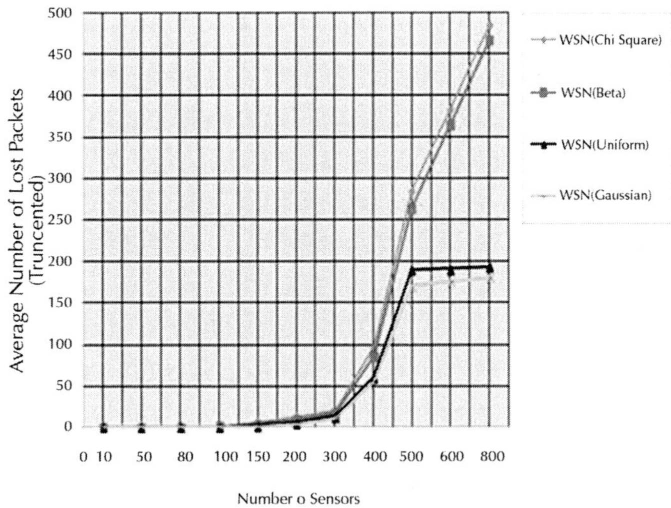

Figure 9: Average end-to-end delay for different WSNs (Gaussian, uniform, beta, and chi-square).

Throughput

Throughput is defined as the correct transmitted data with a specific quality from source to destination within a time interval [40]. Increased number of sensors used in WSNs makes analysis of QoS parameter an important issue. There are many factors affecting the WSN throughput such as packet loss, end-to-end delay, energy consumption, long distance, and obstacles. Figure 10shows the throughput for different WSNs sensors distributed by using Gaussian, uniform, beta, and chi-square distributions. The simulation results indicate that Gaussian WSN achieves the maximum throughput, the second highest throughput is achieved by uniform WSN, and beta and chi-square WSNs have the lowest throughput. The interpretation for these results is that Gaussian and uniform WSN have direct sensor communication and high QoS paths. Furthermore, the WSN task may be shared among multiple sensors due to sensor overlaps as a result of less collisions and packet drops. On the other hand, beta and chi-square WSNs have complex paths as a result of higher collisions and packet drops take place as a result of which the throughput is minimum. Also, energy consumption makes some sensors not to provide their services; so other alternative routes should be generated to complete the routing process.

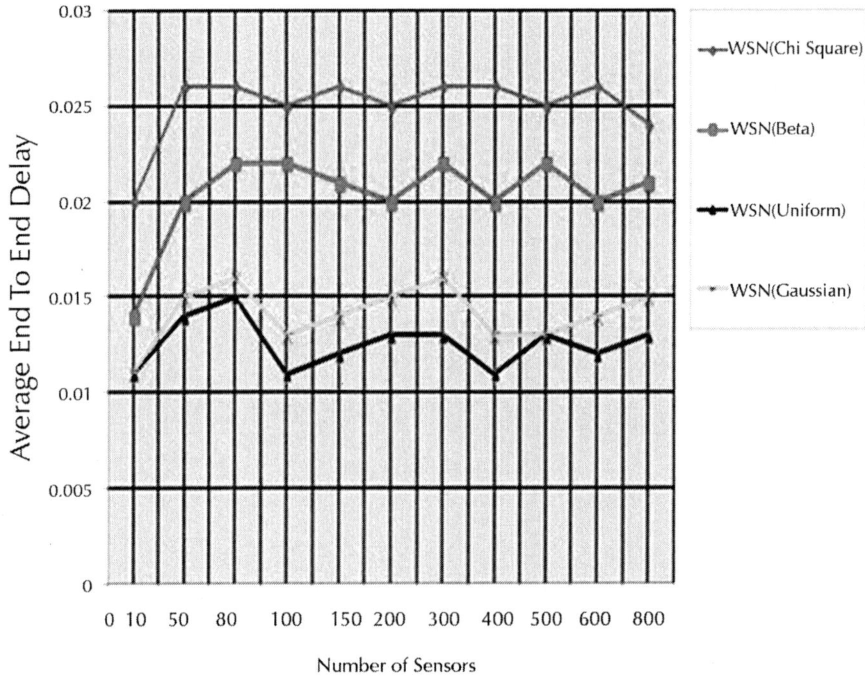

Figure 10: Throughput for different WSNs (Gaussian, uniform, beta, and chi-square).

General Efficiency

The general efficiency of the entire system is defined by a relationship between the numbers of events which are handled by sensors in one WSN in proportion to other WSNs regarding some parameters such as end-to-end delay, packet loss, throughput, and path complexity. The simulation results in Figure 11 show that Gaussian WSN has the best efficiency and then uniform WSN. Beta and chi-square WSNs have the lowest efficiency. The general efficiency parameter is considered as cumulative result which summarizes all the previous results. Also, all previous results and final recommendation are summarized in Table 2.

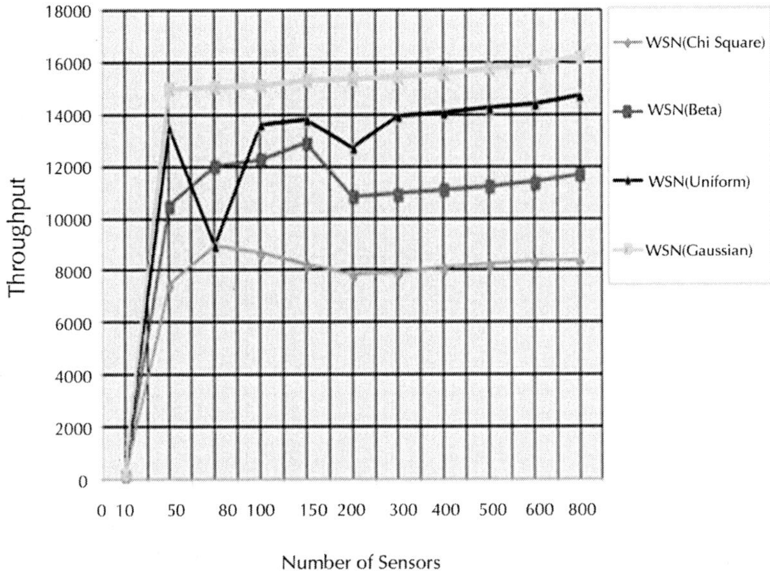

Figure 11: General efficiency for different WSNs (Gaussian, uniform, beta, and chi-square).

Table 2: Results summarization and final recommendation

Distribution	**WSN efficiency evaluation**		**Intruder detection evaluation**		**Final recommendation**
Gaussian	Control data size	HP			
	Number of hops	HP	Without considering distant	HP	
	Packet loss HP				
	End-to-end delay	LPR2			
	Throughput	HP	With considering distant	HP	
	General efficiency	HP			
Uniform	Control data size	LPR2			Using Gaussian
	Number of hops	LPR2	Without considering distant	LPR2	to distribute the
	Packet loss	LPR2			sensors in WSN
	End-to-end delay	HP			gives the best

	Throughput	LPR2	With considering distant	LPR2	performance as
	General efficiency	LPR2			regards intruder detection
Beta	Control data size	LPR3			probability and overall
	Number of hops	LPR3	Without considering distant	LPR3	WSN efficiency,
	Packet loss	LPR3			then uniform, beta,
	End-to-end delay	LPR3			and chi-square
	Throughput	LPR3	With considering distant	LPR3	
	General efficiency	LP			
Chi-square	Control data size	LP			
	Number of hops	LP	Without considering distant	LP	
	Packet loss	LP			
	End-to-end delay	LP			
	Throughput	LP	With considering distant	LP	
	General efficiency	LP			

HP: highest performance. LPR2: lower performance-rank2. LPR3: lower performance-rank3. LP: lowest performance.

Said and Elnashar

Said and Elnashar *EURASIP Journal on Wireless Communications and Networking* 2015 **2015**:46, doi: 10.1186/s13638-015-0240-6

CONCLUSIONS

In this paper, a new model for WSN intruder detection in 3D environment is proposed. Each element in the security system is represented by a center of various radii spheres. Sensors are distributed in the 3D environment using Gaussian, uniform, beta, and chi-square. The security status (hacked or safe) is determined by three parameters, the intruder/target and nearest sensor/target distances, the intruder velocity, and the relationship between spheres (intruder and sensors). The proposed model handles both detection scenarios, single and multiple. Two approaches of evaluations are presented. The first one concerns with WSN security issue. The second approach concerns with network parameters such as average number of control bits, the number of hops, the average number of lost packets, the end-to-end delay, the throughput, and the general efficiency.

The results showed that Gaussian WSN have the best performance in both evaluation approaches, then uniform WSN. The beta and the chi-square WSNs have the lowest performance. Also, the end-to-end delay is the only parameter where the uniform WSN provides better performance than the Gaussian WSN. Simulation results of the proposed model show that Gaussian sensors distribution in WSN is recommended in 3D environments.

Future Work

Mixed probability distributions of sensors in WSN (i.e., Gaussian/uniform, Gaussian/beta, or more mixtures) should be studied and tested. Accordingly, the results should be compared with those of this paper. This will provide a standard sensor distribution. Also, the number of sensors, which are used in the simulation environment, should be larger. In addition, a tracer system for an intruder may be added to the proposed model for system security integrity.

ACKNOWLEDGEMENTS

I would like to pay special thanks and appreciation to the persons who made our research successful.

REFERENCES

1. Y Luo, S Morgera, R Sankar, in proceedings of the 2nd IEEE International Conference on Information Science and Engineering (ICISE). A survey on intrusion detection of wireless sensor network (China, 4–6 Dec 2010), pp. 1798–1802.

2. Akyildiz, Y, Weilian, S, Sankarasubramaniam, Y, Cayirci, E (2002). A survey on wireless sensor networks. *IEEE Commun. Mag., 40*(8), 102–114.
3. K Sohraby, D Minoli, T Znati, Wireless sensor networks: technology, protocols, and applications. Wiley-Interscience, 12–68 (2007).
4. Al-Karaki, J, Kamal, A, Cayirci, E (2004). Routing techniques in wireless sensor networks: a survey. *IEEE Wireless Commun., 11*(6), 6–28.
5. Tilak, S, & Abu-Ghazaleh, N (2002). A taxonomy of wireless micro-sensor network models.*ACM Mobile Comput. Commun. Rev., 6*(2), 28–36.
6. A Agah, S Das, K Basu, M Asadi, in proceedings of the Third IEEE International Symposium on Network Computing and Applications (NCA). Intrusion detection in sensor networks: a non-cooperative game approach (USA, January 30 Aug), pp. 343–346.
7. A Agah, S Das, K Basu, A game theory based approach for security in wireless sensor networks, (USA, Unknown Month 15).
8. Giruka, V, Singhal, M, Royalty, J, Varanasi, S (2008). Security in wireless sensor networks.*Wiley Wireless Commun. Mobile Comput., 8*(1), 1–24.
9. Arora, A, Dutta, P, Bapat, S, Kulathumani, V, Zhang, H, Naik, V, Mittal, V, Cao, H, Demirbas, M, Gouda, M (2004). A line in the sand: a wireless sensor network for target detection, classification, and tracking. *Int. J. Comput. Telecommun. Networking-Special Issue: Mil. Commun. Syst. Technol., 46*(5), 605–634.
10. Solimana, H, Hikalb, N, Sakrb, N (2012). A comparative performance evaluation of intrusion detection techniques for hierarchical wireless sensor networks. *Elsevier Egypt. Inform. J.,13*(3), 225–238.
11. S Sachan, M Wazid, P Singh, H Goudar, A cluster based intrusion detection and prevention technique for misdirection attack inside WSN, (India, May 3).
12. H Kung, D Vlah, Efficient location tracking using sensor networks. Proceedings of IEEE International Conference on Wireless Communications and Networking (WCNC), 1954–1961 (Unknown Month 20).

13. Lin, C, Peng, W, Tseng, Y (2006). Efficient in-network moving object tracking in wireless sensor networks. *IEEE Trans. Mobile Comput.*, *5*(8), 1044–1056.
14. Albagory, Y, & Said, O (2015). Performance enhancement of high-altitude platforms wireless sensor networks using concentric circular arrays. *Elsevier AEU-Int. J. Electronics Commun.*, *69*(1), 382–388.
15. Yang, T, Mu, D, Hu, W, Zhang, H (2014). Energy-efficient border intrusion detection using wireless sensors network. *Springer EURASIP J. Wireless Commun. Netw.*, *1*, 1–12.
16. Mitchell, R, & Chen, I (2014). A survey of intrusion detection in wireless network applications. *Elsevier Comput. Commun. J.*, *42*, 1–23.
17. Lin, C, Chen, W, Tseng, Y (2006). Efficient in-network moving object tracking in wireless sensor networks. *IEEE Trans. Mobile Comput.*, *5*(8), 1044–1056.
18. Y Zhang, W Lee, Intrusion detection in wireless ad-hoc networks, (USA, November 6).
19. Stetsko, A, Smolka, T, Matyas, V (2014). Improving intrusion detection systems for wireless sensor networks. *Springer Appl. Cryptography Netw. Secur. Lect. Notes Comput. Sci.*, *8479*, 343–360.
20. Kur, J, & Matyas, V (2014). A protocol for intrusion detection in location privacy-aware wireless sensor networks. *Springer Trust Privacy Secur. Digital Bus. Lecture Notes Comput. Sci.*, *8647*, 180–190.
21. Y Wang, Y Leow, J Yin, Is straight-line path always the best for intrusion detection in wireless sensor networks, (China, November 8).
22. Y Wang, W Fu, D Agrawal, Intrusion detection in Gaussian distributed heterogeneous wireless sensor networks, (USA, Summer 12).
23. Wang, Y, Wang, X, Wang, B, Agrawal, D (2008). Intrusion detection in homogeneous and heterogeneous wireless sensor networks. *IEEE Trans. Mobile Comput.*, *7*(6), 698–711.
24. Y Wang, F Li, F Fang, Poisson versus Gaussian distribution for object tracking in wireless sensor networks, (China, Unknown Month 22).

25. B Liu, P Brass, O Dousse, P Nain, D Towsley, Mobility improves coverage of sensor networks, (USA, Unknown Month 25).
26. O Dousse, C Tavoularis, P Thiran, Delay of intrusion detection in wireless sensor networks, (Italy, Unknown Month 22).
27. S Qiong, C Comaniciu, Efficient cooperative detection for wireless sentinel networks, (USA, Unknown Month 17).
28. D Turgut, B Turgut, L Boloni, Stealthy dissemination in intruder tracking sensor networks, (Germany, Unknown Month 20).
29. Wang, Y, Fu, W, Agrawal, D (2013). Gaussian versus uniform distribution for intrusion detection in wireless sensor networks. *IEEE Trans. Parallel Distributed Syst.*, *24*(2), 342–355.
30. Chopde, M, Ramteke, K, Kamble, S (2011). Probabilistic model for intrusion detection in wireless sensor network. *Int. J. Commun. Netw. Secur. (IJCNS)*, *1*(3), 19–23.
31. Mubarak, M, Sattar, S, Sajitha, A (2010). Intrusion detection: a probability model for 3D heterogeneous WSN. *Int. J. Comput. Appl.*, *6*(12), 15–20.
32. C Ortiz, J Puig, C Palau, M Esteve, Wireless sensor network modeling and simulation, (Valencia, Spain, Unknown Month 14).
33. Marsaglia, G (1972). Choosing a point from the surface of a sphere. *Ann. Math. Stat.*, *43*(2), 645–646.
34. R Kaparti, OPNET IT GURU: a tool for networking education. MSCIT Practicum Paper. "http://staff.ustc.edu.cn/~bhua/experiments/ITGAE_Tool_Ntwrk_Ed.pdf" Accessed 3 Mar 2015.
35. The network simulator - ns-2. "http://www.isi.edu/nsnam/ns/" Accessed 10 Dec 2014.
36. Xuan, T, Choi, S, Koo, I (2014). A novel blind event detection method for wireless sensor networks. *Hindawi J. Sensors, 2014*, 1–6.
37. D Cohen, M Kelly, X Huang, N Srinath, Trustability based on beta distribution detecting abnormal behaviour nodes in WSN, (Indonesia, Unknown Month 29).
38. Islam, M (2013). Motion analysis using distance and velocity-time function. *Int. J. Sci. Knowl.*, *2*(1), 1–6.
39. Sphere-sphere intersection. "http://mathworld.wolfram.com/Sphere-SphereIntersection.html" Accessed 10 Dec 2014.

40. Akyildiz, I, Sankarasubramaniam, W, Cayirci, E (2002). Wireless sensor networks: a survey. *Elsevier J. Comput. Netw.*, *38*(4), 393–422.

Chapter 5

The Role of Snowmelt Runoff on the Ocean Environment and Scallop Production in Funka Bay, Japan

Satoshi Nakada[1, 6], Katsuhisa Baba[2], Masatoshi Sato[2], Masafumi Natsuike[3], Yoichi Ishikawa[4], Toshiyuki Awaji[5], Koji Koyamada[1], and Sei-Ichi Saitoh[3]

[1]Institute for Liberal Arts and Sciences, Kyoto University, Kitashirakawa-Oiwake Cho, Sakyo-ku, Kyoto 606-8502, Japan

[2]Hokkaido Research Organization, Fisheries Research Department, Hakodate Fisheries Research Institute, 1-2-66 Yunokawa-cho, Hakodate 042-0932, Hokkaido, Japan

[3]Faculty of Fisheries Sciences, Hokkaido University, 3-1-1, Minato-cho, Hakodate 041-8611, Hokkaido, Japan

[4]Data Research Center for Marine-Earth Sciences, Japan Agency for Marine-Earth Science and Technology, Yokohama Institute for Earth Sciences, 3173-25 Showa-machi, Kanazawa-ku, Yokohama 236-0001, Kanagawa, Japan

[5]Kyoto University Headquarters, Yoshida-Honmachi, Sakyo-ku, Kyoto 606-8501, Japan

[6]Graduate School of Maritime Sciences, Kobe University, 5-1-1 Fukae-minami, Higashi-nada-ku, Kobe 658-0022, Japan

ABSTRACT

This study investigated the role of snowmelt runoff on water circulation, water mass formation, and the production of cultured scallop larvae, as a part of a land-sea linkage, by analyzing hydrological data in conjunction with nutrient data and by conducting sensitivity experiments based on a coupled land-sea model of Funka Bay, Japan, a typical semi-enclosed bay. A comparison between observed data and the simulated runoff showed that, using newly compiled datasets of nutrient concentrations in rivers and groundwater, the model was sufficiently accurate to estimate the terrestrial dissolved inorganic nitrogen (DIN) flux from the river and submarine groundwater discharges (SGDs). The average volume flux from the SGDs accounted for 26% of the riverine runoff flux. The DIN flux from SGDs accounted for a maximum of 40% of the total DIN loading to the bay before the snowmelt period. Sensitivity experiments using an ocean simulation indicated that the freshwater flux supplied by snowmelt runoff not only enhances clockwise circulations along with upwelling along the coast, but also modifies the distributions of wintertime water masses in the bay. However, the snowmelt runoff has little effect on larvae transport since wind forcing, rather than riverine buoyancy, dominates the circulation patterns. The annual density of scallop spat was highly correlated with snowmelt runoffs associated with high DIN concentrations, which suggested that riverine nutrients can increase the biomass of phytoplankton in near-shore seas and improve food availability for scallop spawners, resulting in increased egg production in March to April. Therefore, the nutrient flux from agricultural source areas through the large snowmelt runoff has an important role in larvae production. Land-sea linkages need to be identified to design sustainable and synergetic systems of aquaculture and agriculture for the integrated management of coastal regions.

BACKGROUND

Snowmelt runoff can greatly affect not only the ocean's physical environments, through volume and buoyancy fluxes, but also oceanic ecosystems, through nutrient fluxes in coastal seas (Uiboupin and Arino [2010]). Figure 1 displays a map of months showing seasonal river runoff maxima in coastal regions, climatological areas of snow cover, and central lines of intertropical convergence zones (ITCZs). Many rivers located around the ITCZ during June to August had their maximum runoff during the same rainy period, although there were also maximum runoff periods during September and October, with time lags of a few months from the rainy period. The same tendency is predominant for many of the rivers located around the ITCZ during October to December and December to February. Areas of deep snow cover were mainly found in the northern hemisphere and rarely found in the southern hemisphere. The maximum seasonal runoffs occurred between December and August at middle latitudes (35° to 50°N) and between May and October at latitudes higher than 50°N. The maximum seasonal runoffs occurred during March to April and March to June (snowmelt seasons) for middle and higher latitudes, respectively, which could be due to the large-volume snowmelt runoffs that flow into coastal oceans. Therefore, coastal seas, over approximately two-thirds of the northern hemisphere latitude band, can experience snowmelt runoff between spring and early summer.

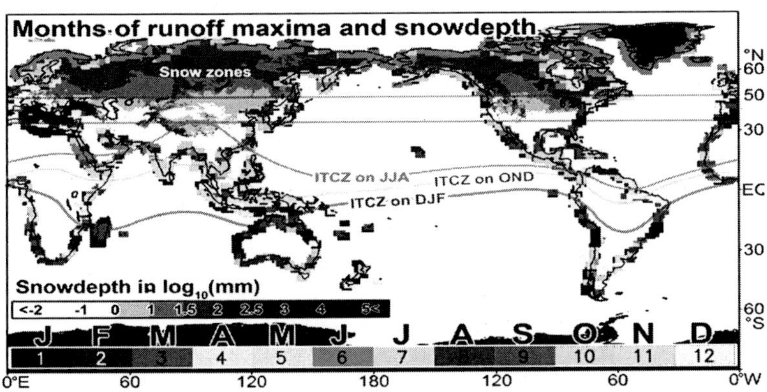

Figure 1: Climatological mean global snow cover and map of months showing seasonal runoff maxima. Chart of climatological mean global snow cover

derived from datasets of the terrestrial water budget data archive (Willmott and Matsuura [2001]) and map of months showing seasonal runoff maxima based on the climatological runoff dataset (Dai and Trenberth [2002]). Red, green, and blue lines indicate the intertropical convergence zones (ITCZ) in JJA (June, July, and August), in OND (October, November, and December), and in DJF (December, January, and February), respectively. Purple lines indicate the latitudes 35°N and 50°N.

Fisheries and aquafarms located in coastal seas can be influenced by terrestrial inputs, such as freshwater, nutrients, and momentum through inflow from rivers (Kimmerer [2002]). Aquacultures operate mainly in coastal oceans and currently contribute approximately 48% of the aquatic animal food for human consumption (FAO [2011]; Bondad-Reantaso et al. [2012]). The fish and shellfish populations of coastal seas are also influenced by the input of terrestrial nutrients through submarine groundwater discharge (SGD) (Liu et al. [2011]; Nakada et al. [2011]). Large snowmelt runoffs, along with high nutrient fluxes, can affect fishery production in coastal oceans. However, no study has yet investigated how snowmelt runoff influences the aquaculture production in coastal oceans. Estimation of nutrient loading through rivers and SGDs around the snowmelt season is essential to investigate the role of snowmelt runoff in fishery production.

A map of the maximum snow depth in Japan (Figure 2) shows that the northern half of Japan is snow-covered during the winter; in particular, the western coastal areas of Honshu and Hokkaido islands are in the snowbelt (maximum snow depth >200 cm). River runoffs from snow-covered areas reached their seasonal maxima between March and May, with the majority of maxima reached in April, excluding local areas around peninsulas and islands (e.g., Noto Peninsula and Sado Island), and the Pacific side. On the Pacific side, runoffs reached maximum levels in April to May and August to September. The relationship between snow-covered areas and runoff maxima indicates that coastal seas may be affected by snowmelt runoffs in Japan. In Funka Bay (northern Japan; Figure 3), the large amount of snowmelt runoff can affect the bay waters, supporting vast primary production through terrestrial and oceanic nutrients (Yoshimura and Kudo [2003]). The Japanese scallop, which represents an important fishery resource in the bay (Minoda and Fuji[1985]), is cultivated in aquafarms (Figure 3) along the coast (10% of the area of Funka Bay). The scallops feed on diatoms to spawn eggs in the spring and during the snowmelt period (Shimada et al. [2000]). Larvae that hatch from the eggs are released from the aquafarms and transported by clockwise circulation (Takahashi et al. [2010]). Aquaculturists must effectively collect larvae or spats released from spawners to achieve high spat densities (Baba et al. [2009]), since the spat density affects the annual production of cultured scallops.

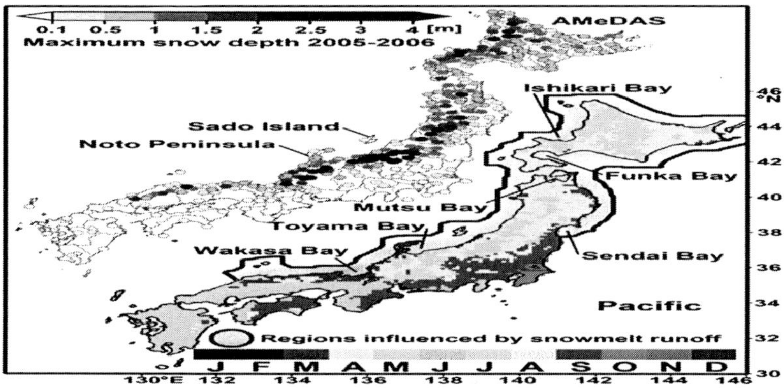

Figure 2: Snow-cover chart and map of months of runoff maxima.Snow-cover chart estimated using the AMeDAS dataset (JMA) and a map of months of runoff maxima in Japan based on simulated data (Yoshimura et al. [2008]). The capital letters along the color bars are abbreviations for each month of the year.

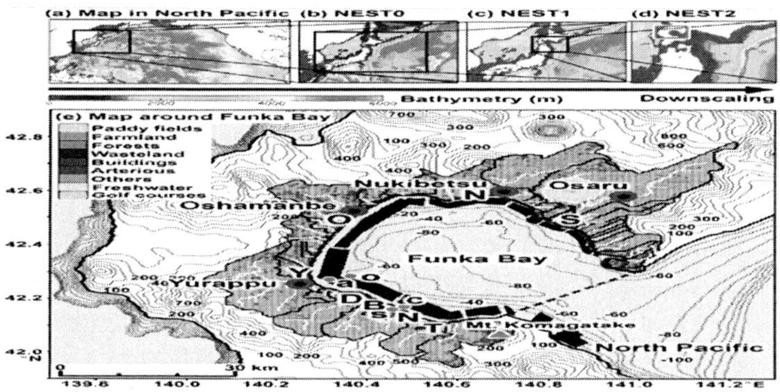

Figure 3: Bathymetry maps of numerical domains in the North Pacific and map of the study area. (a) Bathymetry map of the North Pacific showing the three-step nesting approach from (b) the northwestern North Pacific mod-

el (NEST0) to (c) the medium-scale model (NEST1) to (d) the coastal-scale model (NEST2). Colors indicate the depth in meters. The red square denotes the study area. (e) Map of the study area showing watersheds and aquafarms (black blocks) around Funka Bay. The colors indicate land use in the watersheds. The white lines and colored areas indicate river paths and watersheds, respectively. Blue circles with black dots represent river observation stations with an emphasis on the four major watersheds (solid line). Gray lines indicate isobaths and topographic contours in meters. The dashed line defines the mouth of the bay. The eight capital letters (T, N, B, D, Y, O, N, and S) are the abbreviations for the major rivers (Torizaki, Nigori, Otoshibe, Nodaoi, Yurappu, Oshamanbe, Nukibetsu, and Osaru rivers, respectively). Red dots denote water sampling sites for water in streams (s), near-shore (c), aquafarms (a), and offshore seas (o).

Funka Bay is a typical semi-enclosed bay. The subarctic Oyashio water (OW) and subtropical Tsugaru water (TW), originating from the Kuroshio, alternately flow into Funka Bay from the North Pacific (Figure 4a) and mix over various timescales (Nakada et al. [2013a]). This paper does not distinguish the coastal Oyashio water from the Oyashio water when examining the role of riverine water in the Funka Bay water masses. These unique oceanographic features are crucial for fishery resource management (e.g., skipjack, salmon, and kelp) and ocean biodiversity (Ohtani and Kido[1980]). The majority of rivers that flow into the bay (Figure 3) have small watersheds (up to approximately 500 km^2) with steep gradients (1/25). The mountains surrounding Funka Bay generally grow beech forests (*Fagus crenata*). The watersheds include mostly agricultural lands, but it is unclear whether they are nutrient sources for the primary production around aquafarms. Figure 4b highlights the high concentration of chlorophyll *a* (chl *a*) distributed along the coast, although the presence of colored dissolved organic material (CDOM) and total suspended matter (TSM) in coastal waters may cause errors in chl *a* maps (Darecki and Stramski [2004]). The map of chl *a* implies that terrestrial runoff can supply abundant nutrients and sediments, including CDOM and TSM, around aquafarms through snowmelt events and thereby greatly influences coastal fishery productivity. Terrestrial water can remain in the bay for more than a month owing to weak water exchange (Miyake et al. [1988]).

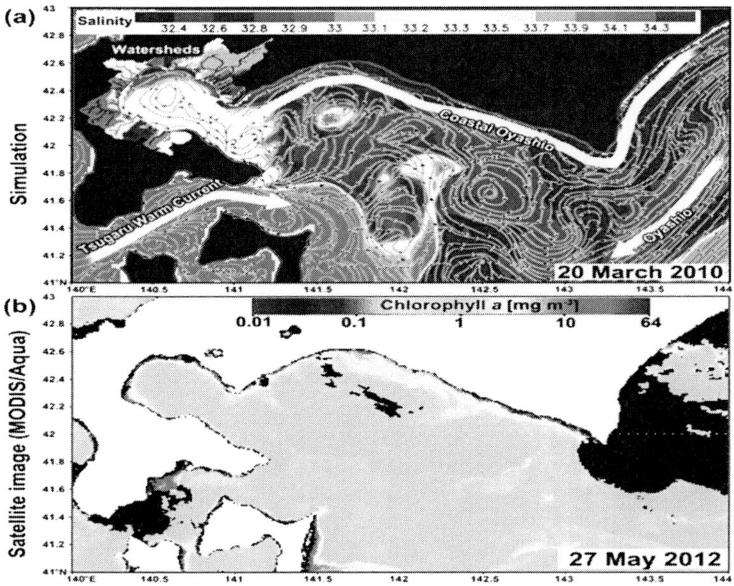

Figure 4: Distributions of simulated salinity and current, and satellite image of chl*a*concentrations. (a) Horizontal distributions of the simulated salinity (colors) and current at a depth of −12 m on 20 March 2010. The white stream lines with arrows indicate the flow direction estimated by the simulated velocity field. (b) Satellite image of chlorophyll *a* concentrations (mg m^{-3}) estimated from MODIS/Aqua ocean color remote sensing data around Funka Bay, southwestern Hokkaido, on 27 May 2012. The land and cloud data are represented by gray and black, respectively.

Recently, a coupled land-sea model was developed to generate an operational ocean prediction system in the bay to inform fishermen of the ocean state in real time (Nakada et al. [2012a]). The model reproduced realistic river runoffs and oceanographic features, such as the clockwise circulation, that agreed quantitatively with observations. New comprehensive analyses can expect to use high-resolution datasets produced by coupled land-sea simulations that are validated by observational datasets. Several studies have employed this approach, using the coupled model to investigate the oceanic dynamical behaviors of sub-mesoscale eddies in coastal oceans (e.g., Zhao et al. [2013]).

In this study, we investigated the role of snowmelt runoff in water circulation, water mass formation, and the production of scallop larvae in the bay to reveal a land-sea linkage using hydrological data in conjunction with nutrient data and by conducting sensitivity experiments based on the coupled land-sea model. The paper is organized as follows: the second section describes the details of numerical methods and nutrient analyses of sampled waters. The third section presents the simulated results and their validations and describes sensitivity experiments for the snowmelt runoff. The fourth section discusses the relationships between the terrestrial runoff and oceanic environment, in terms of nutrient supply and larval transport related to scallop production. The fifth section provides the conclusions.

METHODS

Overview of the Coupled Land-sea Model

To simulate ocean states, land-surface conditions, and river runoff, a coupled land-sea model (Figure 5) was generated using a three-step nesting method with four-dimensional variational data assimilation. The model consisted of the hydrometeorological and multi-runoff utility model (HaRUM) (Nakada et al. [2012a]) and a three-dimensional ocean general circulation model (OGCM) developed at Kyoto University (Ishikawa et al. [2009]). The HaRUM was designed to minimize the calculation cost for ocean forecasts and to calculate the SGD.

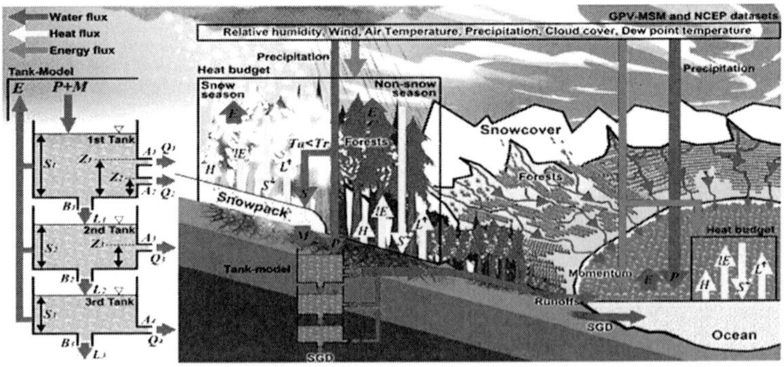

Figure 5: Flowchart to estimate river runoff and structure of the tank model with three serial tanks. The physical quantities depicted in the diagram are

described in the text.

River discharges were predicted using a distributed-tank model, based on radiation, heat, and water mass budgets, by calculating physical components including the solar insolation ($S\downarrow$), downward longwave radiation flux from the atmosphere ($La\downarrow$), upward longwave radiation flux from the ground surface ($Lg\uparrow$), sensible heat flux (H), and latent heat flux (lE), which determines rainfall and snowfall (Tr), precipitation (P), evapotranspiration (E), and snowmelt (M) (see Kondo and Watanabe ([1992]) and Kondo and Yamazaki ([1990]) for detailed formulae). These budgets can be calculated using the five meteorological properties (rainfall, air temperature [Ta], wind speed, cloud cover, and relative humidity) from the grid point value datasets of the meso-scale model (GPV-MSM) provided by the Japan Meteorological Agency on an hourly basis. The river discharge, Qt, generated by the net water input ($P+M-E$) is predicted for each grid of the distributed-tank model (Figure 5). The model has three serial tanks, represented as n = 1 to 3, ten model coefficients (An, Bn, and Sn) for the side outlets (Qn) and bottom outlets (Ln), and specified depth thresholds of water levels (Zn) in each tank. If snow covers the ground surface (Figure 5) and the snow surface temperature satisfies $Ts < Ta$ (where Ts denotes the ground surface temperature), the snowmelt, M, is calculated by a radiation and heat balance.

A three-dimensional OGCM developed at Kyoto University was employed using the hydrostatic and Boussinesq approximations and a realistic bottom topography. The OGCM implemented a hybrid -z vertical coordinate system to improve simulation of the free surface motion of the ocean. To enhance the representation of upper ocean circulation further, this model adopted several sophisticated parameterizations, including a turbulence closure scheme for the mixed layer parameterization (Noh [2004]), UTOPIA (Leonard et al. [1993]) for horizontal advection, and an isopycnal mixing scheme (Gent and McWilliams [1990]; Griffies [1998]). To improve the realistic reproduction of salinity distributions in the coastal ocean or around the aquafarms in Funka Bay, we employed a three-step nesting or downscaling method (Figure 3) that implemented a large-scale, four-dimensional variational data assimilation model (Ishikawa et al. [2009]). The assimilation model (NEST0) covered the western half of the North Pacific at horizontal resolutions of (1/6)° and (1/8)° for longitude and latitude, respectively, and provided accurate boundary conditions for two nested models embedded in the western region. The first nested model (NEST1) without assimilation covered the northwestern North Pacific and centered on the mixed water region at medium resolutions of (1/18)° and (1/24)° for longitude and latitude, respectively. The second nested model (NEST2) without assimilation was resolved with the finest horizontal resolutions of (1/54)° and (1/72)° for longitude and latitude, respectively, to reproduce the small-scale physical features within coastal zones. Note that 78 vertical levels were spaced between 4 m apart near the sea surface and 500 m apart at the bottom. The boundary conditions used in each nested model were provided from upper-level (larger-scale) modeling results using the nesting technique by Oey and Chen ([1992]). Assimilated elements in our study were satellite-derived sea surface temperature, sea surface height data, and *in situ* observation data on temperature and salinity (see Ishikawa et al. ([2009]) for more details). The river runoff derived from the HaRUM was converted to freshwater flux and input into the meshes closest to the coast of Funka Bay. The freshwater flux was reflected by a decrease in salinity as water was elevated by freshwater inputs into the water column at each surface grid. The high-resolution ocean simulation in NEST2 was spun up using initial conditions following the methods of Ishikawa et al. ([2009]).

Reanalysis and predicted meteorological GPV-MSM datasets and the National Centers for Environmental Prediction reanalysis datasets were input into the coupled model using the mean hourly air temperature, precipitation, cloud cover, relative humidity, dew-point temperature, and wind speed. The coupled model was validated using observational data and showed quantitative reproducibility for temporal variations in the bay after several calibrations (Nakada et al. [2012a],[2013a]).

We conducted sensitivity experiments by changing the runoffs in the OGCM to evaluate the influence of snowmelt runoff on circulation and water mass formation in the bay. Two numerical runs were conducted: 1) a case using the terrestrial runoff into the bay as the control run (CR) and 2) a case without the total runoff and SGDs (NF) during spring to summer (March to July).

Water Sampling and Measurements for Nutrient Concentrations

The datasets of nutrient concentrations observed at the mouths of eight major rivers (Figure 3) were used to calculate riverine nutrient fluxes (Yoshimura and Kudo [2003]). To derive further concentration data for the Yurappu River and other streams, additional water samples were collected at river mouths between April and August in 2011 and 2012 (Table 1). In our study, the streams were defined by the small areas of the watersheds (<10 km^2). Seawater was also sampled as a control. The water samples, obtained using acid-cleaned buckets, were collected from the surface (approximately 1 m deep). Nutrients (nitrates plus nitrites) were determined using continuous-flow automated analyzers (QuAAtro, BL TEC K.K., Osaka, Japan; AutoAnalyzer II, Technicon Co., tarry town, New York, the U.S.) following the manufacturers' protocols. The observation dates for the riverine, ground, and stream waters are described in Table 1.

Table 1: Dates of water sampling observations from 2011 to 2013

Year	Date	Water sampling
2011	May 18	River water (Yurappu) and seawater (stations a and o in Figure 3)
	June 20	
	August 29	
	October 11	
2012	March 16	
	April 12	
	April 25	
	May 14	
	June 25	
	August 14	
	December 14	
2013	April 15	Stream water station s, seawater, around station a, c, and o in Figure 3
	April 16	River water (Yurappu)

Nakada et al.

Nakada et al. Progress in Earth and Planetary Science 2014 1:25 doi:10.1186/s40645-014-0025-2

Groundwater data, including nutrient concentrations measured at observational wells, were downloaded from the website of the National Land Information Division, National and Regional Policy Bureau, Ministry of Land, Infrastructure, Transport and Tourism (http://nrb-www.mlit.go.jp/kokjo/inspect/inspect.html), which provided historical observational datasets of groundwater quality. We removed data that lacked information about the elevation at the well or screen depths, because observation depths at the wells could not be calculated without this information. Data without observed water levels were removed as it was not known whether the screens had been submerged in well water. Furthermore, data with observation depths lower than −70 m or higher than 35 m were removed because groundwater located outside these depths is located far from the coast and contributes little to the SGD. A total of 24 (out of 89) wells were chosen for the period of 1980 to 2006.

RESULTS

Simulated Terrestrial Volume Fluxes to the Bay

Figure 6a,b,c,d shows riverine and submarine groundwater discharges based on the HaRUM outputs and runoff observations. Large discharges arising from snowmelt were visible in the simulated and observed runoffs of four major rivers for late April; the discharges had seasonal maxima and displayed interannual variability. The low observed water runoffs associated with the baseflow were reproduced successfully, along with local peaks generated by intense rainfall after the snowmelt period. The temporal tendency after each peak agreed well with the observations and was mainly composed of the interflow discharged from the third outlet. The high degree of consistency between the observed and estimated runoffs for the four major rivers during the observation period resulted in high correlations (approximately 0.8) at significance levels similar to previous studies (Nakada et al. [2012a]). The Nash-Sutcliffe efficiency coefficient (Nash and Sutcliffe [1970]) ranged between 0.6 and 0.7, except for the Yurappu River coefficient (0.4), which fell into the range (0.36 to 0.75) of 'satisfactory' performance (Motovilov et al. [1999]). The relative root-mean-square errors (RMSEs), or the fractional RMSEs, for climatologically averaged runoffs were 21.8% for the Yurappu River (14.5 ± 4.6 $m^3 s^{-1}$), 12.4% for the Osaru River (20.5 ± 7.9 $m^3 s^{-1}$), 31.4% for the Oshamanbe River (7.4 ± 2.9 $m^3 s^{-1}$), and 40.9% for the Nukibetsu River (8.3 ± 3.4 $m^3 s^{-1}$). The average of the fractions was approximately 27%.

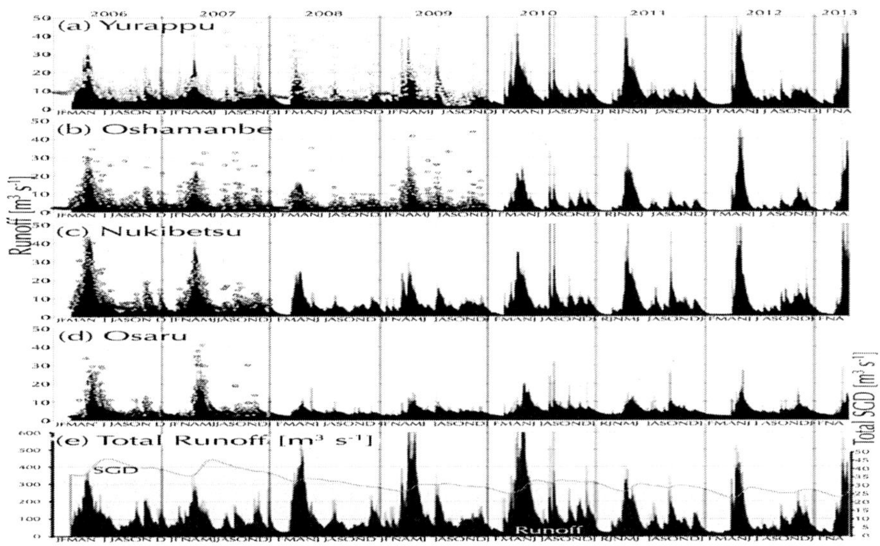

Figure 6: Riverine and submarine groundwater discharges and time series of the simulated volume flux of the SGD. (a-d) Temporal variations in observed (open circles) and simulated (bars) daily runoffs for four major rivers from January 2006 to April 2013. **(e)** Estimated volume transports of the total discharge from all rivers (black) and submarine groundwater discharges (SGDs) (gray). All units are $m^3 s^{-1}$.

The tank model without the SGD effect (L_3; the bottom outlet in the third tank) displayed a positive interannual runoff trend, leading to a decrease in the model's reproducibility (not shown). Meanwhile, the tank model applied for large watersheds (3,000 km^2) did not need the bottom outlet in the lowest tank (e.g., Iwanami et al. [2013]): no SGD condition was required. Watersheds in our study area, which are largely characterized by Neogene or Paleogene volcanic rocks (andesite), retained their high permeability. These facts suggested that the SGD effect is essential for realistic simulations of volume and nutrient fluxes in small watersheds (<500 km^2) with high permeability.

Time series of all runoffs were estimated using the coefficients determined by Nakada et al. ([2012a]) and summed to derive the total discharge from all watersheds. The robust time series of the total discharge (Figure 6e) clearly showed seasonal maxima around late April, in the snowmelt period. The interannual variations of snowmelt runoffs were also visible and indicated that the largest snowmelt runoffs occurred in 2010. The maximum freshwater loading to Funka Bay occurred when the discharge was 300 to 600 $m^3 s^{-1}$ around late April.

Figure 6e shows the time series of the simulated volume flux of the SGD, which indicates marked increases in the flux after the seasonal maxima of snowmelt runoff from the rivers. The seasonal maxima of SGDs were 28 to 46 $m^3 s^{-1}$, occurring around June. The time lag between the seasonal maxima of SGDs and the snowmelt runoff was approximately 1 month. During the snowmelt period, the water level in the third tank seasonally increased owing to the generation of snowmelt water, which led to a flux maximum that was mainly caused by leakage from the third tank. The flux of SGD after the seasonal maximum gradually decreased from summer to fall.

Simulated Terrestrial Nutrient Fluxes to the Bay

The nutrient concentrations of terrestrial waters were estimated based on our field surveys, in addition to the observational dataset of riverine concentrations (Yoshimura and Kudo [2003]). Observed data were mainly derived in spring to early summer (March to July). Figure 7a shows seasonal variations in nutrient concentrations (DIN) for riverine waters sampled at river mouths. Concentrations averaged over all months were used to substitute missing data from summer to winter (August to February), allowing derivation of the time series of the DIN flux from watersheds (except for the Yurappu River). Seasonal peaks in seasonal variations of DIN occurred in March or April, except for the Osaru River. After a seasonal peak, concentrations decreased during the period in April when the snowmelt runoff increased, and became lower than average for each river except the Osaru River. The time lag between DIN and river runoff peaks was approximately 1 month, indicating that nutrient concentrations in rivers were attenuated by the large volume of snowmelt water.

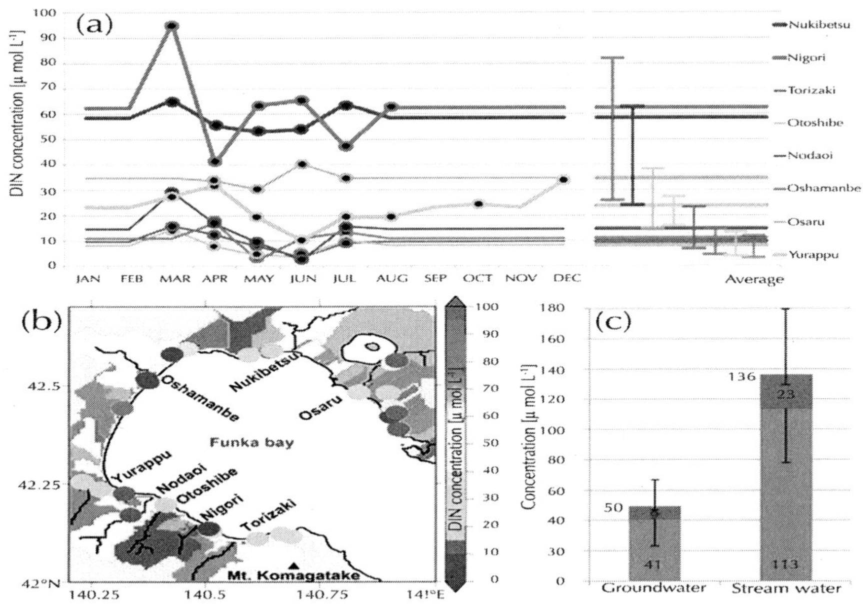

Figure 7: Nutrient concentrations. (a) Seasonal variations (left) and climatological averages (right) of DIN concentrations observed at the mouths of the eight major rivers. Error bars on the averages indicate errors estimated by assuming the nutrient concentration for the missing data period lies between zero and the maximum values. Black dots with each river color indicate values calculated by observational datasets. (b)DIN concentrations of groundwater measured at observation wells. Shaded areas indicate each watershed along the bay. Black lines denote the paths of eight major rivers. (c) The average concentrations of NH_4 (red), $NO_2 + NO_3$ (blue), and DIN (red + blue) for groundwater and stream waters. Error bars indicate the standard errors of the concentrations for NH_4 and $NO_2 + NO_3$ (downward bars) and DIN (upward bars).

It is essential to estimate errors due to lack of data. Lower and upper error limits were estimated by assuming the nutrient concentration for the period in which data were missing to be minimum (zero) and the maximum values, respectively. The lower limit of error bars, with DIN averages, in the right panel in Figure 7a shows that the DIN in the missing data period can be less than half the averaged DIN value, except at the Yurappu River. The climatologically averaged concentration over all riverine waters (28 µmol L^{-1}) includes errors of −20 and +8 µmol L^{-1}. The averaged error value is ±11 µmol L^{-1} or 39% as a fraction of the averaged value. The quantity of data observed at the Yurappu River was greater than for other rivers, suggesting that further observation will reduce these errors. The missing data period did not occur during the snowmelt period, although the error magnitudes were comparable to those for seasonal variations. Therefore, these errors were mostly insensitive to our analyses, but as shown later, errors can influence the climatologically averaged DIN riverine fluxes (Table 2).

Table 2: Climatologically averaged DIN riverine fluxes from the eight major rivers

Rivers	**DIN (mol s^{-1})**	
	Averages	**Errors**
Torizaki	0.04	0.02
Nigori	0.10	0.04
Otoshibe	0.04	0.02
Nodaoi	0.10	0.05
Yurappu	0.37	0.11
Oshamanbe	0.09	0.04
Nukibetsu	0.56	0.28
Osaru	0.74	0.24
Average	0.26	0.14

Nakada et al.

Nakada et al. Progress in Earth and Planetary Science 2014 1:25 doi: 10.1186/s40645-014-0025-2

A total of 24 wells were located near the ocean to monitor groundwater quality, distributed fairly evenly across the coastal areas around the bay (Figure 7b). DIN concentrations in the groundwater from the wells were 1.3 to 418.3 μmol L^{-1}, with an average of 50 ± 18 μmol L^{-1} (Figure 7c), approximately twice the climatologically averaged concentration over all riverine waters (28 ± 8 μmol L^{-1}). A total of 15 sites (63%) showed higher concentrations than the lower limit of the averaged riverine concentrations (20 μmol L^{-1}), and these sites were largely located in or around the watersheds of the Osaru, Yurappu, and Nukibetsu rivers and at the base of Mt. Komagatake. This finding indicated that SGDs from these watersheds can increase the DIN concentration in the coastal bottom water. The average groundwater concentration of NH_4 (9 ± 2 µmol L^{-1}) was much lower than the concentration of $NO_2 + NO_3$ (41 ± 18 µmol L^{-1}). The average concentration of DIN in stream water (NH_4: 23 ± 7 plus NO_2 + NO_3: 113 ± 36 µmol L^{-1}) was 136 ± 43 µmol L^{-1}(approximately five times the concentration in riverine water). High-concentration DIN water could flow directly into the coastal water around aquafarms through streams and SGDs from narrow watersheds (<10 km^2) in the mountains.

Average DIN concentrations in the Nigori and Nukibetsu rivers were two times higher than in other rivers (Figure 7a). These high concentrations could be affected by nutrient runoff from agricultural fields, because paddy fields, farmlands, and golf courses account for 23% to 30% of the land use in the watersheds of both rivers (Figure 8). The fractions of forests in both watersheds ranged from 62% to 74%, lower than the average fraction (75%) for all watersheds. In contrast, agricultural fields accounted for 6% to 17% for the other watersheds, leading to low concentrations of DIN in the associated rivers. Fractions of forests and agricultural fields indicate that nutrient loading from forests does not contribute to the increases in DIN concentrations in rivers. The DIN concentration in the near-shore water, measured around the Nigori River (Figure 3) (4.6 ± 0.2 µmol L^{-1}), was higher than the area-averaged concentration in the aquafarm and near-shore areas (2.0 ± 1.1 µmol L^{-1}) and at offshore observation sites (0.4 ± 0.2 µmol L^{-1}), for April 2011 to 2013.

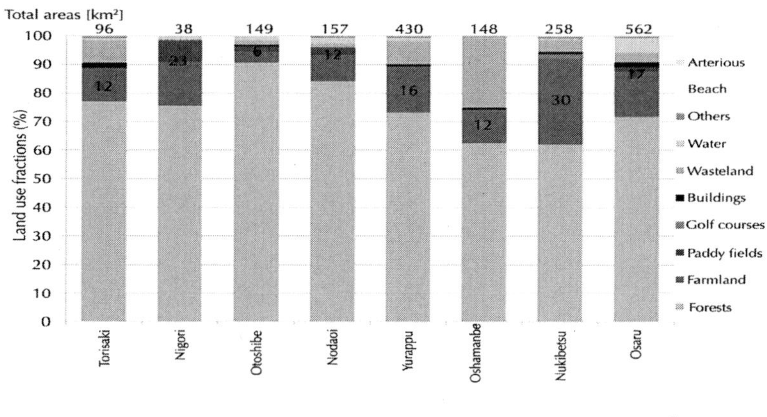

Figure 8: Percentages of land use in the seven major watersheds. See Figure 3 for the map of land use in each watershed. The numerals on the bars rep-

resent the summed percentages of paddy fields, farmlands, and golf courses in the watersheds. The numerals at the top of the bars indicate the areas of each watershed.

The total DIN flux from all rivers and SGDs was calculated by multiplying the river, stream, and submarine groundwater discharges by DIN concentrations in those waters. Figure 9a shows that temporal variations of total DIN fluxes were similar to total volume fluxes for both rivers and SGDs (Figure 6). Variations in volume fluxes determine the variations in DIN fluxes. Since volume fluxes from SGDs were lower than those from rivers, higher DIN concentrations in SGDs had a comparably minor influence on the total DIN flux.

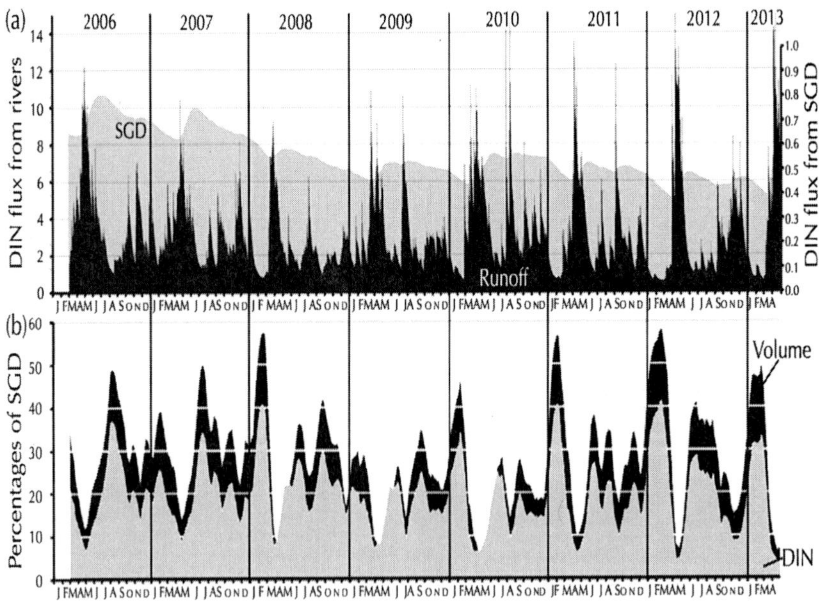

Figure 9: Temporal variations of DIN fluxes. (a) Temporal variations in estimated daily mean DIN fluxes (mol s^{-1}) from all rivers (black) and all submarine groundwater discharges (SGDs) (gray) between January 2006 and April 2013. (b) Temporal variations in percentages of the estimated volume (black) and DIN (gray) fluxes from SGDs divided by the total flux (all fluxes from the rivers plus all fluxes from the SGDs).

Figure 9b shows temporal variations in the percentages of volume and DIN fluxes from SGDs in total fluxes. Variations in the percentages of the volume and DIN were similar, but the percentages of the volume flux were generally higher than those of the DIN flux and often exceeded 50% around February. The percentages of the DIN flux from the SGD in the total terrestrial DIN flux increased from 10% to 15% in the snow-cover season (or winter) and reached its seasonal maximum (30% to 40%) around February for each year except 2007 and 2009. Local maxima also occurred between summer and fall, when both the riverine DIN and volume fluxes were low. Seasonal minima were reached during the snowmelt periods, after seasonal maxima, because of the large DIN flux generated by snowmelt runoff from rivers.

Climatologically averaged DIN fluxes from eight major rivers and their errors (Table 2) were calculated using averaged DIN concentrations (with associated errors) for each river (Figure 7) and averaged volume fluxes and errors (RMSEs), with error calculations based on error propagation. Errors in DIN fluxes from each river ranged from 0.02 to 0.28 mol s^{-1}. The average value was 0.26 ± 0.14 mol s^{-1} or 54% of the averaged value. The error was affected by errors in the averaged DIN concentration (39%), rather than those in the volume flux (27%). The DIN flux from Osaru River was the highest and corresponded to the highest volume flux. In contrast, the Nigori River hardly affected the total DIN flux as it showed the lowest volume flux (1.6 m^3 s^{-1}), although the averaged DIN concentration of the Nigori River was the largest. The Yurappu, Nukibetsu, and Osaru rivers contributed 82% of the total DIN flux because their volume fluxes and DIN concentrations were higher than those for other rivers (Figure 7); however, errors were two to ten times higher than those for other rivers. Therefore, more observations of DIN concentrations in these rivers could decrease errors in future studies.

Figure 10 summarizes climatologically averaged contributions to total volume and DIN fluxes from the major, small (<10 km^2), and other watersheds, corresponding to eight major rivers, streams, and other minor rivers, respectively. Volume fluxes from SGDs accounted for 26% of riverine volume fluxes (123 m^3 s^{-1}). The estimated riverine volume flux accounted for 79% of the total flux (155 m^3 s^{-1}), which was comparable to a previous estimate (150 m^3 s^{-1}; Hasegawa and Isoda [1997]). Volume fluxes from all SGDs (32 m^3 s^{-1}: 21%), and the small and other rivers (37 m^3 s^{-1}: 24%), accounted for 45% of the total flux, which could be a second source of terrestrial freshwater. DIN fluxes from all SGDs accounted for 16% of the riverine DIN flux (3.29 mol s^{-1}), which represents a typical fraction (e.g., Taniguchi et al. [2002]). The calculated DIN flux from the eight major rivers (2.04 mol s^{-1}) was comparable to a previous estimate (2.42 mol s^{-1}; Yoshimura and Kudo [2003]) and accounted for 53% of the total DIN flux (3.84 mol s^{-1}). The remaining 47% is provided by all SGDs (0.55 mol s^{-1}: 14%), and the small and other rivers (1.25 mol s^{-1}: 33%), which indicated that they can supply nutrient fluxes comparable to riverine fluxes from the major rivers. Currently, no study has evaluated these fluxes, but our approach successfully considered each of the fluxes in detail. Streams (narrow watersheds) contributed 5% (6 m^3 s^{-1}) of the riverine volume flux, but contributed 15% (0.50 mol s^{-1}) of the riverine DIN flux owing to high DIN concentrations in the steam water. DIN fluxes of SGDs from small and other watersheds (0.04 and 0.08 mol s^{-1}, respectively) were negligible (1% and 2%, respectively) compared to the total DIN flux. Seasonal differences in each contribution or percentage were rarely observed in volume and DIN fluxes. It is important to note that these respective values can often include errors of (typically) 27% in volume fluxes, an average derived from model errors, and 41% to 54% for the DIN fluxes propagated from model and observational errors.

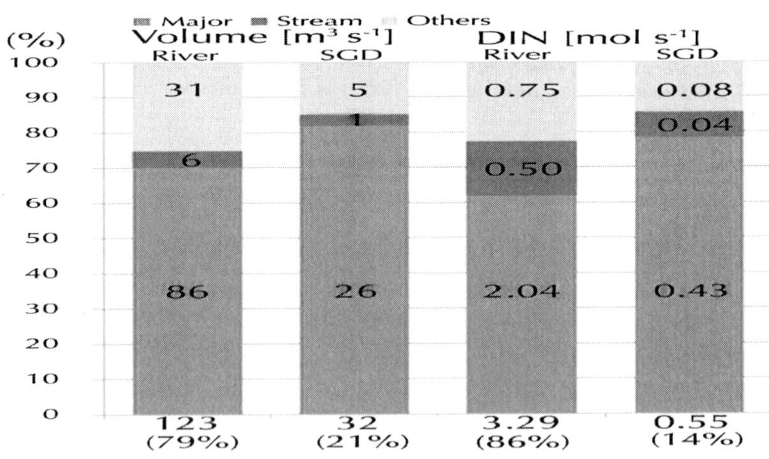

Figure 10: Contributions to total volume and DIN fluxes. Fractions of volume and DIN fluxes from the watershed groups of major rivers (blue), streams (red), and other small rivers (green) averaged over 2006 to 2013. Figures on the color bars denote the climatological volume and DIN fluxes of rivers and SGDs in each watershed group. The values at the bottom of each bar indicate total volume and DIN fluxes from the all rivers and SGDs.

Results Simulated Using the Coupled Land-sea Model

The coupled model correctly simulated the typical ocean phenomena that occurred during spring and summer, such as the Oyashio inflow during the spring (Ohtani [1981]), the summertime clockwise circulation induced by terrestrial freshwater (Satoh et al. [2003]), and the local wind in the bay (Takahashi et al. [2007]). These phenomena were often captured by conductivity-temperature-depth (CTD) sensor and acoustic Doppler current profiler (ADCP) observations (e.g., Takahashi et al. [2010]), which were used to validate the simulated results. The simulated results were validated using observational datasets (salinity and temperature) at satisfactory model skill scores (Murphy and Epstein [1989]) ranging from 0.5 to 0.8 (see Nakada et al. [2013a]).

Figure 11 concurrently shows the horizontal distributions of the simulated salinity, velocity, and terrestrial runoffs. Each distribution of the terrestrial runoff was calculated by multiplying the runoffs by the stream orders at each grid, normalized to the maximum orders in each watershed. We highlight the maps for March and July in 2008 and 2010, which indicate a clear transition from the Oyashio inflow pattern to clockwise circulation. The maps for March (both years) show the anticlockwise circulation generated by the inflow of low-salinity Oyashio water. The large number of visible river paths indicates that large runoffs occurred in March because of snowmelt runoff; their number decreased in July, when predominantly major rivers were found in both years. The low-salinity water distributed along the coast of the bay was formed by snowmelt runoff in March for both years and was most dominant in July. Low-salinity water originating from terrestrial discharge in July formed around the estuaries of the major rivers, indicating the regions of freshwater influence (Simpson [1997]). Similar anticlockwise and clockwise circulations also formed during other years (not shown).

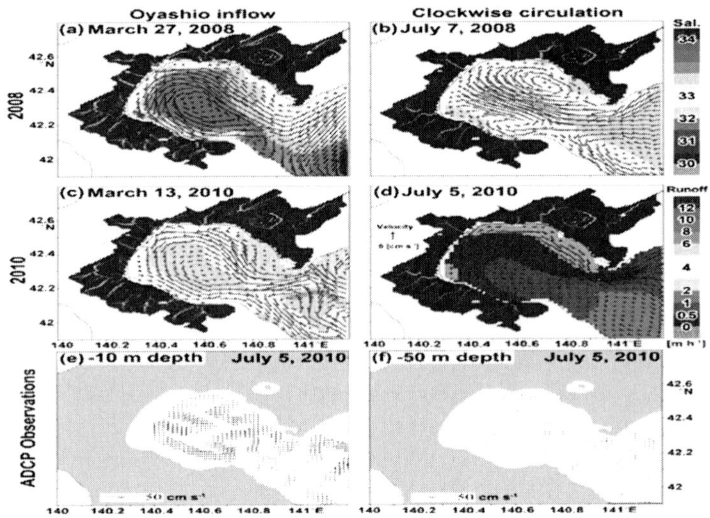

Figure 11: Horizontal distributions of the simulated salinity, velocity, and terrestrial runoffs. Daily mean maps of the simulated salinity (colors on the ocean), current (vectors) at the surface layer (−2 m depth), and terrestrial runoff (colors on the land) (a) on 27 March 2008, (b) on 7 July 2008, (c) on 13 March 2010, and (d) on 5 July 2010. The levels of runoff volume transport are depicted by red to blue to white colors along the river paths. The lower panels show snapshots of the observed velocity field at depths of (e) –5 m and (f) –50 m on 5 July 2010. Velocity fields were measured using the acoustic Doppler current profiler installed on the R/V *Kinsei-maru* of the HFRI.

The observed velocity field at the surface (−10 m depth) on July 5 (Figure 11e) shows a clockwise circulation that is similar to the simulation, although current speeds for the circulation were generally five times greater than simulated speeds. The underestimation of the surface velocity field can be attributed to the model's vertical resolution; the layer thickness around the surface layer, from 0 to 20 m, is coarse (4 m), which may insufficiently resolve the surface boundary layer. Local inconsistencies between the simulation and observations were found to correspond to detailed flow patterns around the bay mouth and along the northern coast in the outer bay. The inconsistency around the bay mouth can be attributed to tidal components, for example, the 10 cm s^{-1} velocity observed at the bay mouth (Ohtani [1981]) can be attributed to the M2 tidal component. The local wind pattern can induce the observed inconsistency in the outer bay, because the horizontal resolution of the wind (GPV/MSM) was originally 5 km, insufficient to resolve the 10 km spatial pattern of the wind. The observed speed at a depth of −50 m (Figure 11f) was comparable to the simulated speeds (not shown), which also suggests that the simulation underestimated the surface velocity field.

Sensitivity Experiments for Snowmelt Runoff

Influence of Runoff on Circulation

Sensitivity experiments were conducted for 1 year, 2008, because this year showed typical snowmelt runoffs (Figure 6) and oceanographic events, including a typical Oyashio and Tsugaru warm water inflow (Figure 11). The salinity and velocity fields, averaged over March to April, in the control run (CR) case (upper panels of Figure 12) represented the average pattern of the Oyashio inflow that is associated with a low-salinity flow, forming the anticlockwise velocity field. The low-salinity areas formed along the coast of the bay and were less saline during the snowmelt period (March to April) than during early summer (May to June). The velocity field, averaged over May to June, represented the large clockwise circulation.

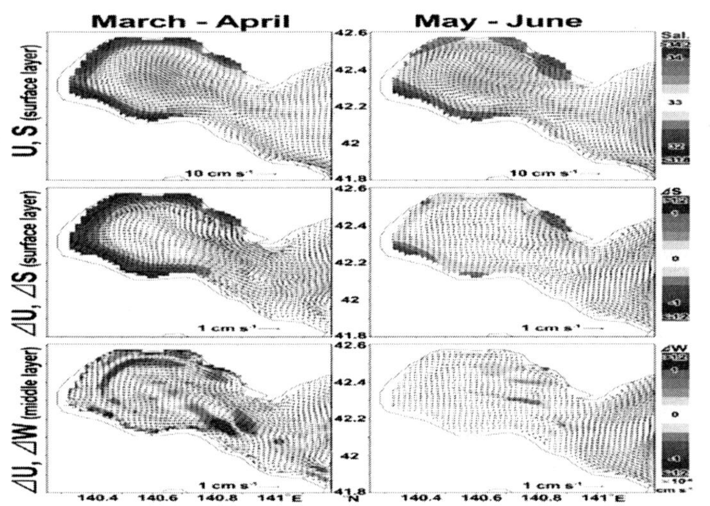

Figure 12: Bi-monthly mean maps. Upper panels: bi-monthly mean maps of the simulated salinity (colors) and current (vectors) at the surface layer (−2 m

depth) in (left) March to April and (right) May to June for the case including terrestrial runoff. Middle panels: bi-monthly mean maps of the differences in simulated salinity and current at the surface layer between the cases with and without terrestrial runoff, over the same period as the upper panels. Lower panels: differences in the vertical velocity (colors) and horizontal current (vectors) averaged over the middle layer (depths from −10 to −30 m) between the cases with and without terrestrial runoff, over the same period as the upper panels.

The difference between surface salinity maps for the CR case and the NF case, which is without terrestrial runoffs during March to July (middle panels of Figure 12), indicated that the terrestrial discharge associated with snowmelt decreased the salinity field (by approximately −1.1) along the coast. The salinity difference became small during May to June; the areas of marked salinity difference (less than −0.6) in May to June were confined to the areas around major river estuaries.

Differences in surface velocity fields during the period of snowmelt showed that offshore flows from rivers formed along the coast. These seaward flows deflected to the left, leading to a clockwise velocity field around the center of the bay. Maps of the vertical velocity difference (lower panels of Figure 12) at depths from −10 to −30 m showed upwelling in the near-shore area along the coast of the bay and downwelling in the offshore area. The flow pattern characterized by seaward flows and upwelling can be regarded as the estuarine circulation along the coast of the bay. This arrangement suggests that the terrestrial discharge driven by snowmelt could generate the estuarine circulation along the entire coast of the bay. The clockwise velocity field, enhanced by snowmelt runoff, is significant in March to April; however, the estuarine circulation diminished in May to June, and the clockwise flows weakened and moved closer to the coast.

The temporal variation of the vorticity difference between the CR and NF cases (CR minus NF) (Figure 13) showed that the vorticity difference decreased from early March to the beginning of April during the snowmelt period. The sign of the vorticity difference changed from positive to negative, indicating the transition process from the anticlockwise flow pattern associated with the Oyashio inflow to clockwise circulation in the bay. The decrease in the vorticity difference was highly correlated with the terrestrial discharge (correlation coefficient = 0.92), with a time lag of 10 days. The time lag may indicate the period needed to reach the stable negative vorticity generated by buoyancy due to the terrestrial runoff through the geostrophic adjustment. Once the clockwise velocity field was established until the peak of the snowmelt runoff in the middle of April, the negative vorticity was retained by the small runoff although the clockwise velocity field gradually weakened because of the spin down. Our sensitivity experiment clearly showed that the terrestrial buoyancy flux enhanced the clockwise velocity field associated with estuarine circulation.

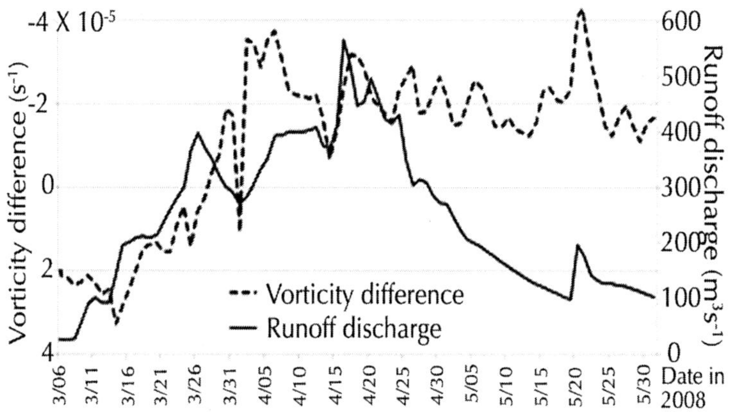

Figure 13: Temporal variation of the vorticity difference between the CR and NF cases. Temporal variation in the daily mean difference of the simulated vorticity (CR minus NF) averaged over the surface layer (−2 m depth) of the bay (dashed line), and estimated daily mean volume flux from the terrestrial runoff (black line) from early March to late May 2008.

Influence of Runoff on Water Mass Formation

The experiment indicated that the low-salinity Funka Bay surface water (FS) was reduced in the NF case (without snowmelt runoffs) compared with the CR case (Figure 14a). The FS was predominantly formed by riverine water flowing into the bay rather than rainfall on the bay. The mixed water (MW), which is a mixture of the FS, the summertime OW (S-OW), and the Tsugaru Water (TW) largely formed at depths of −50 to −70 m in both cases. The amount of MW increased from June to November in the CR case (Figure 14b). In contrast, the amount of MW decreased from October in the NF case and was much smaller than for the CR case. The lower salinity water formed by terrestrial freshwater is mixed with the TW by vertical convection from the beginning of the surface cooling period in September through the winter, leading to the emergent MW in winter. The terrestrial freshwater input yields wintertime MW that is less saline than the TW, and can suppress the increase of the TW volume in the CR case, and salinization of the bay water. The TW, at a depth of −40 to −80 m, is an apparent intrusion due to the emergent MW. The bottom intrusions in August, September, and November were visible at depths below −80 m, flushing out the oxygen-deficient bottom water and suppressing the mortality rate of the cultured scallop. Meanwhile, the volume of the TW in the NF case continuously increased from September. The wintertime TW in the CR case ranged from 4×10^{10} to 9×10^{10} m^3. However, the amount of wintertime TW (approximately 14×10^{10} m^3) in the NF case was 1.4 to 3 times that in the CR case, where it occupied almost all of the bay water. The amount of OW in the NF case (maximum volume of 4×10^{10} m^3) was 1.2 times that in the CR case (maximum volume of 3.5×10^{10} m^3), which suggests that the clockwise vorticity, enhanced by terrestrial discharge, can suppress the anticlockwise inflow of the Oyashio water (Figure 12). Highly saline Funka Bay water formed during the winter (FW) in the NF case (1.9×10^{10} m^3); it was also 1.2 times higher than for the CR case (1.6×10^{10} m^3) during spring of the following year (2009; not shown).

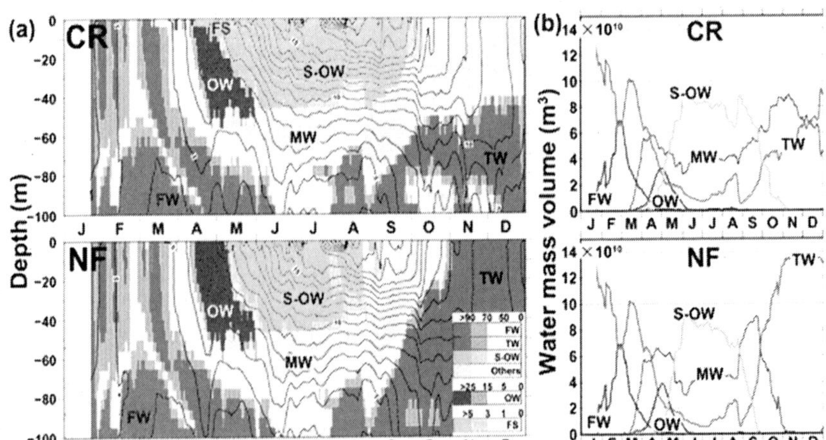

Figure 14: Seasonal and temporal variations of water masses. (a)Seasonal variation in the vertical distribution of water masses and area-averaged simu-

lated temperatures over Funka Bay. The temperature contour interval is 1°C. The letters indicate the six water masses: the Tsugaru warm current water system (TW), the Oyashio water system (OW), the warm low-salinity Funka Bay water that formed during the summer (FS), the cold dense Funka Bay water that formed during the winter (FW) (Hanawa and Mitsudera [1986]; Ohtani [1981]), the OW surface waters that were warmed mainly by sea-surface heating (S-OW; Rosa et al.[2007]), and all other water, which is defined as mixed water (MW). These water masses were classified using the T-S diagram for this study (see Figure Three of Nakada et al.[2013a]). The colors indicate percentages of each water mass occupying the water volume at each depth. (b) Temporal variations of water mass volumes in the bay.

DISCUSSION

Relationship between Terrestrial Discharges, Nutrients, and Larval Production

After the winter cooling, nutrients in the bay are vertically homogeneous in February (Lee et al.[2002]). A massive spring phytoplankton bloom, characterized by a high chl *a* concentration (>20 μg L^{-1}), is then induced by high-nutrient waters, increasing solar radiation, and water column stability during March to April (Kudo and Matsunaga [1999]) and is terminated by nitrate (NO_3) depletion (Kudo et al. [2000]). The riverine nutrients can contribute to the summertime primary production in the entire bay (Yoshimura and Kudo [2003]). This study found that terrestrial nutrient loading in the coastal ocean is driven predominantly not only by riverine flux, but also by the nutrient flux through the SGD that accounts for a maximum of 40% of the total flux before the snowmelt period. This DIN flux also contributed to the primary production in near-shore seas.

Our results revealed that DIN concentrations were higher in the near-shore and aquafarm areas than in offshore observation sites during the snowmelt period. The concentration of chl *a* in near-shore seas (Figure 4) is strongly associated with terrestrial nutrients. The concentrations of riverine nutrients were much higher than concentrations in the mixed layer of the bay (9 to 12 μmol L^{-1}), even before the spring bloom (Kudo et al. [2000]). This suggests that the high DIN flux from rivers can significantly contribute to the increase in DIN concentration, primary production, and chl *a* in the shallow coastal seas during the winter.

The *in situ* chl *a* concentration in February is highly correlated with the density of scallop spat because the scallop spawners can improve food availability in years of high chl *a* and allocate enough energy to gonadal development (Baba et al. [2009]). The correlation between the simulated volume flux and the observed spat density (Figure 15) was also highest in spring for each year except 2011. The remarkably high density of the spat in 2011 (Figure 15) might have been influenced by the Great East Japan Earthquake on March 11, because the tsunami that surged through the coastal area of the bay disrupted the sea floor sediment, leading to high concentrations of nutrients in the surface layer. Higher correlations were generally found in March to April than in February to April, which suggests that the terrestrial nutrient flux generated by the large snowmelt runoff in March to April played a more significant role in food availability than the flux in February. Spawners utilized the food or chl *a* produced by the terrestrial nutrients in March to April before the scallops' spawning season from late April until late May (Kosaka and Ito[2006]). The eggs and larvae can float in the surface ocean without feeding on phytoplankton until they become young scallops and attach to aquaculture facilities for approximately 40 days. Therefore, eggs and larvae should not be dependent on the phytoplankton supplied by the bloom or the Oyashio inflow in March to April. The nutrients or chl *a* affected by oceanic events could have a significant role in food availability for the young scallops feeding on the phytoplankton.

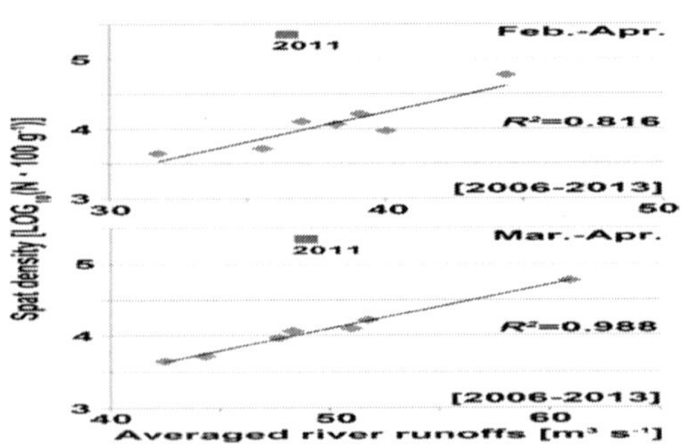

Figure 15: Scatter plots of spat densities. For scallops and simulated volume fluxes averaged over February to April and March to April of 2006 to 2013

(see Baba et al. [2009] for the sampling and calculation procedures used to derive the spat density data). R^2 denotes the squared correlation coefficient.

The late inflow of cold Oyashio water, from late April through to May, can decrease the egg and larval production because the low temperature stunts gonad growth (e.g., Sastry and BlakeSource[1971]). Sensitivity experiments showed that the clockwise velocity modified by terrestrial discharge could diminish the anticlockwise inflow of cold Oyashio water (Figure 14), which can suppress the decrease of the water temperature in the bay. The high correlation between the simulated volume flux and the observed spat density (Figure 15) can secondarily explain how the relationship between larval production and Oyashio water volume is affected by the terrestrial discharge.

The observed chl *a* in February around the aquafarm is considered to be an index of the food availability for scallop spawners and is used for yearly predictions of spat density (Baba et al.[2009]). The 2013 prediction, using the observed chl *a*, was inconsistent with the observed spat density because the observed chl *a* can reflect spatial and temporal variations, or inhomogeneity, of chl *a* concentration affected by intermittent inputs of both terrestrial and oceanic nutrients through the biochemical mechanism of phytoplankton (Yoshimura and Kudo [2011]). The prediction's representativeness may be limited by the location of the observation site. This study emphasized that predictions of spat density based on chl *a* concentration could be improved by including information about terrestrial runoff or nutrient flux, instead of examining the space-time inhomogeneity of chl *a*.

DIN nutrient concentrations from the Nigori and Nukibetsu rivers were twice the climatologically averaged concentration of all riverine waters (28 ± 8 μmol L^{-1}), owing to the supply from agricultural lands in the watersheds. The proportion of agricultural areas in these watersheds determines the baseline concentration of terrestrial nutrients (Ileva et al. [2009]). The nutrient concentration during the snowmelt period was diluted by the large volume of snowmelt water. However, the temporal variation in the nutrient flux was controlled by the volume flux rather than the nutrient concentration. Similar results have been suggested by previous studies (Iwanami et al. [2013]). This tendency for dilution during the snowmelt period may be a general feature of snow-covered coastal areas around the subarctic zone. The spawning density was affected by riverine nutrients in near-shore areas, which suggests that nutrient loading from agricultural lands played an essential role in food availability for spawners. However, additional studies are needed to determine whether the current loading of terrestrial nutrients is appropriate for scallop aquaculture. The increase in the nutrient flux can improve food availability, but it can also induce harmful algal blooms of organisms such as *Alexandrium tamarense* (Anderson et al. [2012]).

Water Circulation and Larval Migration

The large snowmelt runoff induces upwelling and seaward advection as components of estuarine circulation (Figure 12) and forms nutrient-rich low-salinity water in the surface layer of Funka Bay. The rich nutrients that upwell from the deeper layer may contribute to high chl *a* concentrations near the aquafarms. Terrestrial water accumulates in the bay because the clockwise circulation, driven by the local wind (Takahashi et al. [2010]) and terrestrial runoff (Satoh et al. [2003]), can restrict the exchange of bay water with external water (Isoda et al. [2000]), resulting in a low-salinity surface water (Hasegawa and Isoda [1997]). After the snowmelt period, larvae can be trapped along the coast and transported to the northern coast by clockwise circulation.

The clockwise circulation can be referred to as a type 2 buoyant plume (Garvine [2001]), as the buoyant water at the river mouth turns left along the coast in the northern hemisphere. This water can be trapped on the coast and self-advected along the left-hand coast (Chapman and Lentz[1994]). Satoh et al. ([2003]) interpreted the clockwise circulation as a geostrophic flow proceeding along the left-handed coast, induced by an internal Kelvin-wave propagation under the condition of a preexisting oceanic mixed layer, based on the theory proposed by McCreary et al. ([1997]). Sensitivity experiments on freshwater inputs in the ocean model indicated that the buoyancy flux supplied by the river runoff contributed secondarily to the velocity field or the clockwise circulation in the bay. The contribution of the riverine buoyancy to the circulation was estimated at approximately 20% (= Vr(1 [cm s^{-1}])/Vs(5 [cm s^{-1}]) × 100) of the total momentum of the clockwise velocity field, where Vr is the typical velocity difference that is revealed by sensitivity experiments or induced by riverine runoffs (Figure 12) and V_s is the typical simulated velocity of the clockwise circulation (Figure 11); the remaining 80% was presumably supplied by the wind (Takahashi et al.

[2007]). Episodically, strong winds can form sub-mesoscale vortices in the bay and distort the clockwise circulation (Nakada et al. [2012b]). Such vortices have often been observed in other coastal seas and can locally modify ocean currents (Nakada et al. [2013b]).

CONCLUSIONS

This study comprehensively investigated the roles of snowmelt runoff in water circulation in terms of riverine buoyancy flux, in water mass formation in terms of freshwater volume flux, and in the production of scallop larvae in terms of riverine nutrient flux, for Funka Bay, based on hydrological and nutrient data. Sensitivity experiments using a high-resolution coupled land-sea simulation were also conducted. The terrestrial freshwater and nutrient (DIN) fluxes from rivers and SGDs into the bay were estimated using a hydrological model and climatological datasets compiled from nutrient concentration data for the rivers, streams, and groundwater. The comparison between observed and simulated runoffs indicated that the model accurately reproduced the terrestrial runoff, allowing for estimates of the terrestrial nutrient flux because variations in terrestrial volume fluxes controlled variations in the terrestrial DIN fluxes. The average volume flux from SGDs accounted for 26% of the flux for riverine runoff. The nutrient flux from SGDs accounted for a maximum of 40% of the total loading to the coastal ocean of the bay in February before the snowmelt period.

The impact of snowmelt runoff on water circulation and water mass formation in the bay was first evaluated by sensitivity experiments using the OGCM. This study revealed, as part of the land-sea linkage, that snowmelt runoff has three important roles in the ocean environment and scallop production: 1) Buoyancy fluxes supplied from the runoff can enhance the clockwise circulation in the bay, but rarely dominate the velocity field. 2) The terrestrial freshwater input can diminish the springtime inflow of the cold OW, leading to an increase in larval production and a decrease in the volume of wintertime TW in the bay, and can modify distributions of the wintertime water masses. 3) Fluxes of high-concentration nutrients from the runoff can form a favorable environment for spawners by improving food availability. Therefore, the snowmelt runoff hardly contributes to larvae transport but can increase egg production through the spring nutrient supply. In addition, upwelling may advect rich nutrients from deeper regions to the surface in the near-shore ocean, and the clockwise circulation may transport larvae northward along the coast.

Agricultural fields were important sources of nutrients for feeding scallop spawners. This study, therefore, demonstrated that the integrated management of agricultural farms and aquafarms (Iwasaki [2013]) in the large bay is essential for controlling the nutrient flux from the land to the ocean. The volume, buoyancy, and nutrient fluxes supplied from terrestrial snowmelt runoff can be key factors in fishery or aquaculture production in subarctic coastal regions, alongside watersheds in the snow zone. Additional studies are needed to elucidate land-ocean linkages and to design sustainable and synergetic systems of aquaculture and agriculture.

AUTHORS' CONTRIBUTIONS

SN designed this work; collected the data; carried out the water sampling; conducted the numerical simulation; analyzed, observed, and simulated the data; and prepared the manuscript. KB observed the larvae and spat density and helped in their interpretation. MS observed physical properties such as water temperature, salinity, and current velocity. MN carried out the water sampling and assayed the water to derive nutrient concentrations. YI prepared the ocean general circulation model and its data assimilation system. KK provided the high-end computational environment, visualization system, and super-computing resources. TA and SS proposed the topic, provided discussion, and collaborated with the corresponding author in the construction of the manuscript. All authors read and approved the final manuscript.

ACKNOWLEDGEMENTS

We thank Mr. M. Watanobe, Mr. M. Kanamori, and the crew and officers of the R/V Kinsei-Maru of Hakodate Fisheries Research Institute for providing *in situ* data. We deeply thank Professor I. Kudo (Faculty of Fisheries Sciences, Hokkaido University) and Dr. T. Yoshimura (Central Research Institute of Electric Power Industry) for suggesting improvements to this paper and providing the nutrient datasets (Yoshimura and Kudo [2003]), which greatly helped in the calculation of the terrestrial DIN flux. This work was partially supported by the 'Hakodate Marine Bio Cluster Project' in the knowledge Cluster Program from 2009, the Grant-in-Aid for University and Society Collaboration from the Ministry of Education, Culture, Sports, Science and Technology (MEXT), Japan, and the Grant-in-Aid for Research Programs on Climate Change Adaptation (RECCA). We acknowledge the National Aeronautics and Space Administration (NASA) for the MODIS/Aqua data and the Research Institute for Sustainable Humanosphere, Kyoto University, for the archived data of GPV-MSM. We deeply appreciate the constructive comments from two anonymous reviewers and editors.

REFERENCES

1. Anderson DM, Alpermann TJ, Cembella AD, Collos Y, Masseret E, Montresor M (2012) The globally distributed genus *Alexandrium*: multifaceted roles in marine ecosystems and impacts on human health. Harmful Algae 14:10-35
2. Baba K, Sugawara R, Nitta H, Endou K, Miyazono A (2009) Relationship between spat density, food availability, and growth of spawners in cultured *Mizuhopecten yessoensis* in Funka Bay: concurrence with El Nino southern oscillation. Can J Fish Aquat Sci 66:6-17 doi:0.1139/F08-183
3. Bondad-Reantaso MG, Subasinghe RP, Josupeit H, Cai J, Zhou X (2012) The role of crustacean fisheries and aquaculture in global food security: past, present and future. J Invert Path 110(2):158-165 doi:org/10.1016/j.jip.2012.03.010
4. Chapman DC, Lentz SJ (1994) Trapping of a coastal density front by the bottom boundary layer. J Phys Oceanogr 24:1464-1479 doi: http://dx.doi.org/10.1175/1520-0485(1994)024<1464:TOACDF>2.0.CO;2
5. Dai A, Trenberth KE (2002) Estimates of freshwater discharge from continents: latitudinal and seasonal variations. J Hydro 3:660-87 doi:10.1175/1525-7541(2002)003<0660:EOFDFC>2.0.CO;2
6. Darecki M, Stramski D (2004) An evaluation of MODIS and SeaWiFS bio-optical algorithms in the Baltic Sea. Remote Sens Environ 89(3):326-350 doi:10.1016/j.rse.2003.10.012

7. (2011) FAO yearbook: fishery and aquaculture statistics 2009. Food and Agriculture Organization of the United Nations, Rome, Italy.
8. Garvine RW (2001) The impact of model configuration in studies of buoyant coastal discharge. J Mar Res 59:193-225 doi:10.1357/002224001762882637
9. Gent PR, McWilliams JC (1990) Isopycnal mixing in the ocean circulation model. J Phys Oceanogr 20:150-155 doi:10.1175/1520-0485(1990)020<0150:IMIOCM>2.0.CO;2
10. Griffies SM (1998) The Gent-McWilliams skew flux. J Phys Oceanogr 28:831-841 doi:10.1175/1520-0485(1998)028<0831:TGMSF>2.0.CO;2
11. Hanawa K, Mitsudera H (1986) Variations of water system distribution in the Sanriku coastal area. J Oceanogr Soc Japan 42:435-446
12. Hasegawa N, Isoda Y (1997) Fresh water budget of Funka Bay. Umi to Sora 73:113-121 (in Japanese with English abstract)
13. Ileva NY, Shibata H, Satoh F, Sasa K, Ueda H (2009) Relationship between the riverine nitrate-nitrogen concentration and the land use in the Teshio River watershed, North Japan. Sustain Sci 4:189-198 doi:10.1007/s11625-009-0081-5
14. Ishikawa Y, Awaji T, Toyoda T, In T, Nishina K, Nakayama T, Shima S, Masuda S (2009) High-resolution synthetic monitoring by a 4-dimensional variational data assimilation system in the northwestern North Pacific. J Mar Syst 78:237-248 doi:10.1016/j.jmarsys.2009.02.016
15. Isoda Y, Takeyama Y, Azumaya T (2000) Ocean structural change around the diatom bloom in Funka Bay. Bull Coast Oceanogr 38:3-14.
16. Iwanami H, Wada T, Sakamoto K, Kudo I, Chikita KA (2013) Mechanism of nutrients supply from the Tokachi River catchment considering characteristics of discharge and land use. J Jpn Assoc Hydro Sci 43(1):3-24
17. Iwasaki S (2013) Fishers-based watershed management in Lake Saroma, Japan. Ocean & Coastal Manag 81:58-65
18. Kimmerer WJ (2002) Effects of freshwater flow on abundance of estuarine organisms: physical effects or trophic linkages? Mar Eco Prog Series 243:39-55

19. Kondo J, Watanabe T (1992) Studies on the bulk transfer coefficients over a vegetated surface with a multilayer energy budget model. J Atmo Sci 49(23):2183-2199 doi:10.1175/1520-0469(1992)049<2183:SOTBTC>2.0.CO;2

20. Kondo J, Yamazaki T (1990) A prediction model for snowmelt, snow surface temperature and freezing depth using a heat balance method. J Appl Meteorol 29:375-384 doi:10.1175/1520-0450(1990)029<0375:APMFSS>2.0.CO;2

21. Kosaka Y, Ito H (2006) Japan. In: Shumway SE, Parsons GJ (eds) Scallops: biology, ecology and aquaculture, Elsevier, Amsterdam, the Netherlands. pp 1093-1141

22. Kudo I, Matsunaga K (1999) Environmental factors affecting the occurrence and production of the spring phytoplankton bloom in Funka Bay, Japan. J Oceanogr 55:505-513

23. Kudo I, Yoshimura T, Yanada M, Matsunaga K (2000) Exhaustion of nitrate terminates a phytoplankton bloom in Funka Bay, Japan: change in SiO_4:NO_3 consumption rate during the bloom. Mar Ecol Prog Ser 193:45-51

24. Lee CW, Kudo I, Yokokawa T, Yanada M, Maita Y (2002) Dynamics of bacterial respiration and related growth efficiency, dissolved nutrients and dissolved oxygen concentration in a subarctic coastal embayment. Mar Freshw Res 53:1-7

25. Leonard BP, MacVean MK, Lock AP (1993) Positivity-preserving numerical schemes for multidimensional advection. NASA Tech Memo 106055: 62 ICOMP-93-05, Washington, DC.

26. Liu SM, Li RH, Zhang GL, Wang DR, Du JZ, Herbeck LS (2011) The impact of anthropogenic activities on nutrient dynamics in the tropical Wenchanghe and Wenjiaohe estuary and lagoon system in East Hainan, China. Mar Chemist 125:49-68

27. McCreary JP, Zhang S, Shetye SR (1997) Coastal circulation driven by river outflow in a variable-density 1 1/2-layer model. J Geophys Res 102:15535-15554

28. Minoda T, Fuji A (1985) Biological environments in Funka Bay. Tokai University Press p 1106, (In Japanese), Coastal Oceanography of Japanese Islands.

29. Miyake H, Tanka I, Murakami T (1988) Outflow of water from Funka Bay, Hokkaido, during early spring. J Oceanogr Soc Japan 44:163-170 doi:10.1007/BF02302640

30. Motovilov Y, Gottschalk GL, England K, Rodhe A (1999) Validation of distributed hydrological model against spatial observations. Agric Forest Meteo 98–99:257-277 doi:10.1016/S0168- 1923(99)00102-1

31. Murphy AH, Epstein ES (1989) Skill scores and correlation coefficients in model verification. Mon Weather Rev 117:572-581 doi:10.1175/1520-0493(1989)117<0572:SSACCI>2.0.CO;2

32. Nakada S, Yasumoto J, Taniguchi M, Ishitobi T (2011) Submarine groundwater discharge and seawater circulation in a subterranean estuarine beneath a tidal flat. Hydrol Process 25:2755-2763 doi:10.1002/hyp.8016

33. Nakada S, Ishikawa Y, Awaji T, In T, Shima S, Nakayama T, Isada T, Saitoh S (2012) Modeling runoff into a region of freshwater influence for improved ocean prediction: an application in Funka Bay. Hydro Res Lett 6:47-52

34. Nakada S, Uenaka T, Ishikawa Y, Matsui K, Sakamoto N, Koyamada K, Awaji T, Saitoh S (2012) A visualization study of vortexes extracted from the tremendous ocean simulation data: an application in Funka Bay. Jpn Soc Sim Tech 4(4):145-152 doi:org/10.1016/j.pocean.2013.10.008

35. Nakada S, Ishikawa Y, Awaji T, In T, Koyamada K, Watanobe M, Okumura H, Nishida Y, Saitoh S (2013) An integrated approach to the heat and water mass dynamics of a large bay: high-resolution simulations of Funka Bay, Japan. J Geophys Res Ocean 118:1-18 doi:10.1002/jgrc.20262

36. Nakada S, Hirose N, Senjyu T, Fukudome K, Tsuji T, Okei N (2013) Operational ocean prediction experiments for smart coastal fishing. Prog Oceanogr 121:125-140 doi.org/10.1016/j.pocean.2013.10.008

37. Nash JE, Sutcliffe JV (1970) River flow forecasting through conceptual models part I - a discussion of principles. J Hydro 10(3):282-290 doi:10.1016/0022-1694(70)90255-6

38. Noh Y (2004) Sensitivity to wave breaking and the Prandtl number in the ocean mixed layer model and its dependence on latitude. Geophys Res Lett 31:L23305 doi:10.1029/2004GL021289

39. Oey LY, Chen P (1992) A nested-grid ocean model - with application to the simulation of meanders and eddies in the Norwegian coastal current. J Geophys Res 97:20063-20086
40. Ohtani K (1981) Funka Bay (I) physical environments. Bull Coastal Oceanogr 19:68-80 (in Japanese)
41. Ohtani K, Kido K (1980) Oceanographic structure in Funka Bay. Bull Fac Fish Hokkaido University, Hakodate, Japan.
42. Rosa AL, Isoda Y, Uehara K, Aiki T (2007) Seasonal variations of water system distribution and flow patterns in the southern sea area of Hokkaido, Japan. J Oceanogr 63:573-588 doi:10.1007/s10872-007-0051-4
43. Sastry AM, BlakeSource NJ (1971) Regulation of gonad development in the bay scallop,*Aequipecten irradians*. Lamarck Biol Bull 140(2):274-283
44. Satoh C, Isoda Y, Shimizu M (2003) Clockwise circulation in the upper layer of Funka Bay in summer. Bull Coast Oceanogr 40:181-188
45. Shimada H, Nishida Y, Ito Y, Mizushima T (2000) Relationship among growth and survival of cultured scallops (*Patinopecten yessoensis* JAY), and environmental condition in the coastal area off Yakumo, Funka Bay, Hokkaido, Japan. Sci Rep Hokkaido Fish Exp Stn 27:49-62
46. Simpson JH (1997) Physical processes in the ROFI regime. J Mar Syst 12:3-15 doi:10.1016/S0924-7963(96)00085-1
47. Takahashi D, Kido K, Nishida Y, Kobayashi N, Higaki N, Miyake H (2007) Dynamical structure and wind-driven upwelling in a summertime anticyclonic eddy within Funka Bay, Hokkaido, Japan. Conti Shelf Res 27:1928-1946
48. Takahashi D, Miyake H, Nakayama T, Kobayashi N, Kido K, Nishida Y (2010) Response of a summertime anticyclonic eddy to wind forcing in Funka Bay, Hokkaido, Japan. Conti Shelf Res 30:1435-1449 doi:10.1016/j.csr.2010.05.003
49. Taniguchi M, Burnett WC, Cable JE, Turner JV (2002) Investigations of submarine groundwater discharge. Hydrol Process 16:2115-2129 doi:10.1002/hyp.1145
50. Uiboupin R, Arino O (2010) Study of snowmelt impact on SST and TSM fields in the coastal zone of Barents Sea. IEEE International

Geoscience and Remote Sensing Symposium, pp 4212–4215, doi:10.1109/IGARSS.2010.5654375

51. [http://climate.geog.udel.edu/~climate/html_pages/archive.html] Willmott CJ, Matsuura K (2001) Terrestrial water budget data archive: monthly time series (1950-1999).

52. Yoshimura T, Kudo I (2003) Riverine nutrient loadings and their impact on primary production in Funka Bay, Japan. Oceanogr Jpn 12:185-193

53. Yoshimura T, Kudo I (2011) Seasonal phosphorus depletion and microbial responses to the change in phosphorus, availability in a subarctic coastal environment. Mar Chemi 126:182-192

54. Yoshimura K, Sakimura T, Oki T, Kanae S, Seto S (2008) Toward flood risk prediction: a statistical approach using a 29-year river discharge simulation over Japan. Hydro Res Lett 2:22-26 doi:10.3178/hrl.2.22

55. Zhao K, Nakada S, Sakamoto N, Koyamada K, Bajaj C, Ishikawa Y, Awaji T, In T, Saitoh S (2013) A visualization for the dynamic behaviors of the mixture of water mass for Northwestern Pacific near Japan. Int J Model Simul Sci Comput 4(1341002):18 doi:10.1142/S179396231341002X

Chapter 6

The Role of Chitin, Chitinases, and Chitinase-like Proteins in Pediatric Lung Diseases

Ines Mack[1], Andreas Hector[2], Marlene Ballbach[2], Julius Kohlhäufl[2], Katharina J Fuchs[3], Alexander Weber[3], Marcus A Mall[4], and Dominik Hartl[2]

[1]Department of Pediatrics/UKBB, University of Basel, Petersplatz 1, Basel, 4003, Switzerland

[2]Children's Hospital, University of Tübingen, Hoppe-Seyler-Strasse 1, Tübingen, 72076, Germany

[3]Interfaculty Institute for Cell Biology, Department of Immunology, University of Tübingen, Geschwister-Scholl-Platz, Tübingen, 72074, Germany

[4]Department of Translational Pulmonology, Division of Pediatric Pulmonology and Allergy and Cystic Fibrosis Center, Translational Lung Research Center Heidelberg (TLRC), Member of the German Center for Lung Research (DZL), University of Heidelberg, Grabengasse 1, Heidelberg, 69117, Germany

ABSTRACT

Chitin, after cellulose, the second most abundant biopolymer on earth, is a key component of insects, fungi, and house-dust mites. Lower life forms are endowed with chitinases to defend themselves against chitin-bearing pathogens. Unexpectedly, humans were also found to express chitinases as well as chitinase-like proteins that modulate immune responses. Particularly, increased levels of the chitinase-like protein YKL-40 have been associated with severe asthma, cystic fibrosis, and other inflammatory disease conditions. Here, we summarize and discuss the potential role of chitin, chitinases, and chitinase-like proteins in pediatric lung diseases.

INTRODUCTION

The role of chitin and chitinases has been firmly established in the field of plant and microbial immunity by demonstrating that host-derived chitinases cleave chitin to protect against invading chitin-bearing pathogens, such as fungi. Although mammals lack endogenous chitin or chitin synthases, chitinases and chitinase-like proteins are endogenously expressed in their lung and other organs. Particularly, chitinase-like proteins have been described as dysregulated in a variety of diseases characterized by chronic inflammation and tissue remodeling, yet their potential role for humans has just recently begun to evolve [1],[2]. Chitin is a major component of a variety of allergy-triggering environmental components, including house-dust mites or fungal spores, and fungal asthma is increasingly appreciated as an under-diagnosed disease entity [3]. Thus, an understanding of the complex immunological and pathophysiological implications of chitin-chitinase interactions in the human body is of high relevance for identifying new biomarkers and therapeutic targets for fungal diseases and other conditions, where chitin-coated microbial derivatives play a critical role. Here, we provide an overview of an emerging, yet complex field of research. Of particular interest are interspecies differences with resulting specific nomenclatures. Subsequent to an overall introduction of chitin, the role of chitinases and chitinase-like proteins in pediatric lung diseases are reviewed, leading up to a summary of ideas how these mechanisms could be exploited to improve diagnostics and therapeutics in lung diseases in childhood and beyond (Figure 1).

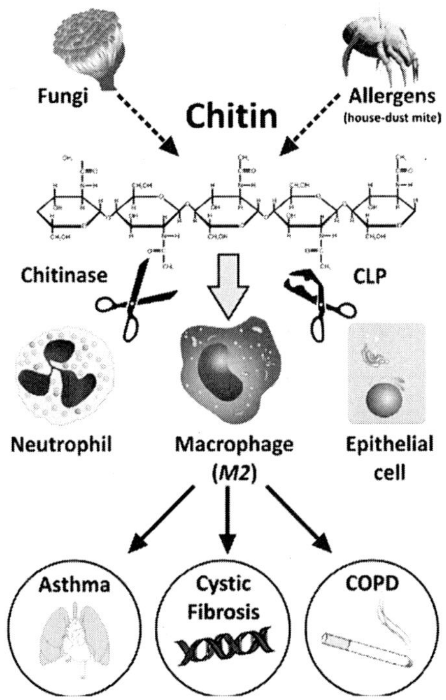

Figure 1: The proposed role of chitin, chitinases and chitinase-like proteins (CLPs) in lung diseases. Chitin is a common component of allergy-triggering environmental components, including fungal spores and house-dust mites, which trigger an innate immune response, including chitinases (cleaving chitin; scissors) and chitinase-like proteins (binding, but not cleaving chitin; damaged scissors). Chitinases and chitinase-like proteins are mainly secreted by neutrophils, alternatively activated macrophages (M2 macrophages) and epithelial cells. The interplay of M2 macrophages, neutrophils, and epithelial cells drives inflammation and remodeling in chronic lung diseases, particularly asthma, cystic fibrosis, and COPD.

REVIEW

Chitin

Chitin, a polymer of *N*-acetylglucosamine and the second most abundant polysaccharide in nature following cellulose, is an essential component of fungi, house-dust mites, exoskeletons of crabs, shrimp and insects, parasitic nematodes, and digestive tracts of many insects [1]. Chitin protects these microbes from their environment, and its turnover is regulated by biosynthesis and degradation through endogenous chitinases.

The first immune stimulatory activity of chitin and chitin derivatives in mammals was discovered and extensively explored in the middle to late 1980s, as reviewed recently [2],[1]. These early studies clearly indicated that chitin has important immunologic effects *in vitro* and *in vivo*, initially highlighted by Shibata et al. who demonstrated that chitin activates peritoneal/alveolar macrophages and natural killer (NK) cells to express a number of pro-inflammatory cytokines such as interleukin-1 (IL-1), colony-stimulating factor (CSF), and gamma interferon (IFN-) [4]. More recent studies by Reese et al. addressed the *in vivo* immune effects of chitin in mice [5]. They noted that after several hours of chitin exposure, innate immune cells were recruited to the lung and/or the peritoneum. These studies also demonstrated that chitin induced alternative macrophage activation and that macrophage depletion with clodronate liposome treatment prevented the recruitment of eosinophils. Van Dyken et al. further demonstrated that fungal chitin from asthma-associated home environments induced allergic eosinophilic lung inflammation [6]. Collectively, these studies strongly suggest that chitin can contribute to the development of allergic type 2 (Th2) inflammation by activating innate immune cells. Beyond these findings, chitin has also been proposed to serve as an immunoadjuvant. In murine asthma models, Shibata et al. demonstrated that orally administered chitin suppressed the production of Th2 cytokines and IgE [7] and, when used as an adjuvant, chitin caused similar effects as a Th1-promoting adjuvant [8],[9]. Accordingly, one can speculate that the immune system senses chitin and that chitin can skew the Th1/Th2 immune response in a bidirectional way.

The studies noted above suggest that the size of the chitin fragment is a crucial determinant of the effector responses that it elicits. This can be seen in comparisons of large chitin polymers that are biologically inert and intermediately sized fragments, which trigger IL-17, IL-23, and TNF- production [1], while even smaller fragments enhance the production of the anti-inflammatory cytokine IL-10. Therefore, chitin has size-dependent effects on murine immune cell function and may bind to different receptors in a size-dependent manner, similar to findings in plants [10],[11]. The observation that large chitin polymers are inert, smaller fragments are pro-inflammatory, and even smaller fragments exert an anti-inflammatory effect allows for an interesting hypothesis regarding the importance of the size-dependent effects of chitin in this response. However, in all of these studies, it is difficult to rule out that chitin preparations contained mixtures of differently long chitin polymers or that 'large' chitin preparations were containing contaminating 'smaller' chitin fragments. Clearly, a more thorough study using well-defined chitin fragments is warranted. Collectively, it is tempting to speculate that chitin recognition by pattern recognition receptors triggers the induction of chitinases, leading to the generation of small-sized chitin particles that are taken up by the host cell. Recently, fungal chitin was found to dampen inflammation through IL-10 induction mediated by activation of the intracellular receptors NOD2 and TLR9 [12]. Despite these intriguing insights, the precise molecular recognition principles of chitin perception remain incompletely understood. Interestingly, this hypothesis is very similar to the established biology of another polysaccharide, hyaluronin that also serves as an alarm signal after degradation. Thus, the ability of appropriately sized polysaccharides to induce inflammation may be a more general principle of glycobiology [2],[1]. When viewed in combination, chitin is a central component of potential pro-allergic microbes (e.g., *Aspergillus fumigatus* and house-dust mite) and has been shown to drive Th2-associated immune responses. Mechanisms that interfere with chitin metabolism are therefore of high relevance for allergic diseases and infections with chitin-bearing pathogens such as fungi.

Chitinases and Chitinase-like Proteins

Chitin-degrading enzymes, known as chitinases, are produced by humans and other mammals and are part of the 18-glycosyl-hydrolase family that encompasses both enzymatically active chitinases and chitinase-like proteins, the latter also termed chi-lectins, which lack enzymatic activity. In humans, acidic mammalian chitinase (AMCase), chitotriosidase, oviductin, and human cartilage glycoprotein (HcGP)-39/YKL-40 and YKL-39 have been described, while YM-1, YM-2, AMCase, oviductin, and breast regression protein (BRP-39) have been identified in mice [13],[1],[14]-[17]. Humans express two functional chitinases, chitotriosidase (*CHIT1*) and AMCase (*CHIA*) with an acidic pH optimum, both able to degrade chitin polymers. In mammals, chitinases [18],[19] and chitinase-like proteins [20] are mainly expressed and secreted by phagocytes (mainly neutrophils and macrophages) and are induced at sites of inflammation, infection, and tissue remodeling, suggesting that these proteins play active roles in anti-infective defense and repair responses. Specifically, both chitinases and chitinase-like proteins have been linked to an alternative activation (M2) phenotype of macrophages [21],[1],[22], which is found in asthma and other chronic diseases, such as cystic fibrosis (CF), providing a rationale for chitinases and chitinase-like proteins to play a role in these disease conditions. While chitotriosidase is used as disease biomarker for Gaucher disease, a disease characterized by the accumulation of lipid-laden macrophages, acidic mammalian chitinase has been linked to allergic asthma and hypersensitivities [23]. However, the major evidence for an involvement in lung diseases exists for chitinase-like proteins. Therefore, we will focus on their role in pediatric lung diseases in the chapters below.

The Chitinase-like Protein YKL-40 and its Involvement in Lung Diseases

YKL-40, or also called HcGP-39 or BRP-39 in mice, lacks measurable enzymatic chitinase activity, due to mutations in its highly conserved putative active sites [17]. In mice, BRP-39 has been associated with cell growth, breast cancer, and tissue remodeling. Recent evidence from BRP-39 knock-out mice indicates that BRP-39 plays a role in T-cell, macrophage and dendritic cell responses as well as cell apoptosis and tissue repair [21],[1]. In human cells, HcGP-39/YKL-40 has been shown to regulate apoptosis/proliferation/cell survival, MAPK, and cytokine pathways [24],[2].

The first indication that YKL-40 could be linked to human lung diseases came from a multi-center study quantifying YKL-40 serum levels in 253 adult patients with asthma [25]. This study showed that YKL-40 serum levels were mainly increased in adult patients with severe asthma and correlated with disease severity and airway remodeling. These studies in adults were, on the one hand, confirmed for children with severe, therapy-resistant asthma [26], but, on the other hand, challenged by another study [27]. Besides increased levels of YKL-40, elevated *bona fide* chitinase activities were also found in bronchoalveolar-lavage (BAL) fluids from children with asthma [28]. YKL-40 was also found to be increased in BAL fluid after segmental allergen challenge, indicating local production of this chitinase-like protein in response to allergens [29]. Gavala et al. confirmed this finding and further demonstrated that segmental allergen challenge also increased chitinase activities [30]. Clinically, YKL-40 serum levels remained increased in patients in spite of long-term inhaled corticosteroids, which could imply that YKL-40 production is resistant to current asthma treatments and might represent an alternative therapeutic target for severe asthma. YKL-40 is mainly released by activated neutrophils [20], and neutrophilic asthma is well known to be corticosteroid-resistant [25]. Thus, increased YKL-40 levels may be a hallmark of neutrophilic asthma. A follow-up study analyzing single nucleotide polymorphisms (SNPs) in the *YKL-40/CHI3L1*gene showed a genetic association with increased susceptibility to asthma, increased bronchial hyperresponsiveness, and reduced lung function [31]. The role of *YKL-40/CHI3L1* SNPs for asthma was further confirmed in a Taiwanese population [32]. Another study found a different *YKL-40/CHI3L1* variant associated with asthma [33]. Given the enigmatic role of YKL-40, the functional importance of these variants awaits further investigation. Moreover, a Chinese study found that YKL-40 levels were increased in asthmatic patients and correlated with exacerbation, eosinophils, and immunoglobulin E [34], while a study from Poland found increased YKL-40 levels in asthma, but no correlation with disease severity or total IgE levels [35]. Besides a positive correlation between YKL-40 levels and age in subjects with asthma across all age groups, Santos et al. found no difference in circulating YKL-40 levels among asthma severities in children nor a correlation with IgE levels [27]. When viewed in combination, some but not all studies support the notion that YKL-40 is increased in severe and/or neutrophilic asthma, yet the clear relationship with age/disease progression (pediatric vs adult) and atopy remains to be defined in future studies.

Beyond asthma, other pulmonary or lung-associated disease conditions where YKL-40 levels were found to be increased were chronic obstructive pulmonary disease (COPD) [36],[37], idiopathic pulmonary fibrosis [38], tuberculous pleural effusions and pneumonia [39], small-cell lung cancer[40], non-small-cell lung cancer [41], bronchiolitis obliterans syndrome [42], hyperoxic acute lung injury [43], sarcoidosis [44], allergic rhinitis [45], and CF lung disease [46]. Despite the heterogeneity of these disease conditions, alternative (M2) macrophage activation is a common feature of the majority of them, suggesting that increased YKL-40 levels reflect M2 macrophage polarization and disease conditions featuring M2 activation [24],[22],[2],[47].

In CF, a chronic neutrophilic inflammatory disease [48],[49], YKL-40 BALF levels were found to reflect airway inflammation and infection in early CF lung disease [50] and correlated inversely with lung function in adult CF patients, where YKL-40 levels were also found to be increased systemically [46]. The potential role of YKL-40 for the pathogenesis of CF lung disease is further supported by findings in *Scnn1b*-transgenic mice, a murine model of CF-like lung disease [51],[52]. In this model, airway-specific overexpression of the -subunit of the epithelial Na^+ channel ENaC mimics airway surface dehydration characteristic for CF airways and produces a CF-like lung disease with early onset of mucus obstruction, chronic airway inflammation, slowed bacterial clearance, and progressive structural lung damage [53]. Similar to patients with CF, levels of the murine homologue of YKL-40, BRP-39, were significantly increased in BALF and showed an inverse correlation with pulmonary function in *Scnn1b*-transgenic mice [46]. Collectively, these studies suggest that YKL-40/BRP-39 may be implicated in the pathogenesis of chronic airway inflammation and airflow obstruction and thus serve as a potential biomarker of disease severity in patients with CF.

Recent studies in this mouse model of CF-like lung disease also provided first mechanistic insights on how chitinases/chitinase-like proteins (CLPs), including BRP-39, may be upregulated in CF lung disease. These studies demonstrated that CF-like airway surface dehydration causing mucociliary dysfunction and mucus obstruction [54],[55] provides a robust stimulus for macrophage activation, even when *Scnn1b*-transgenic mice are kept in a germ-free environment [54], [56]. More recently, a series of gene expression studies identified signatures of alternatively activated macrophages (M2), including *Ym1, Ym2,* and *BRP-39,* in whole lungs and isolated macrophages from *Scnn1b*-transgenic mice [57], [58]. These results suggest alternative macrophage activation in mucostatic airways as a mechanism underlying elevated expression of a range of chitinases and CLPs, even in the absence of chitin-containing parasites or allergens, in the airways of patients with CF and potentially other muco-obstructive lung diseases. While the pathogenic role of chitinases/CLPs in (dys) regulation of inflammation in CF airways remains poorly understood, observations in lungs from *Scnn1b*-transgenic mice demonstrate that some proteins of the chitinase/CLP family, such as*Ym1* and *Ym2*, are expressed at such high levels that they precipitate and form sharp crystals often greater than 100 μm in size [54]. These may cause chronic mechanical irritation and injury of airway epithelial cells and phagocytes thus contributing to chronic airway inflammation. Additionally, it is intriguing to speculate that such crystals may activate the NLRP3 inflammasome, a pattern recognition receptor able to trigger inflammation in response to other crystalline or aggregated endogenous substances such as cholesterol or uric acid crystals [59],[60],[61].

Beyond asthma and CF, circulating YKL-40 has been further associated with decline of lung function in the general population and has been proposed as a biomarker of susceptibility to the long-term effects of cigarette smoking [62]. Studying chitinases in New York City firefighters after World Trade Center exposure revealed that increased serum chitotriosidase reduced the odds of developing pulmonary obstruction after World Trade Center-particulate matter exposure and was associated with recovery of lung function. The underlying mechanisms remain unclear [63]. As chitin is a structural component of fungi, which are well known for their role in environmental asthma, a study investigated whether the exposure to environmental fungi modulates the effect of chitinases in individuals with asthma. The study demonstrated that environmental exposure to fungi modified the effect of *CHIT1* SNPs on severe asthma exacerbations [64].

In order to dissect the cellular sources of YKL-40 in human airways and the mechanisms regulating YKL-40 expression, Park and coworkers identified human airway epithelial cells as a source of YKL-40 and demonstrated that mechanical stress potently induces CHI3L1 expression leading to increased secretion of YKL-40 protein in an EGFR and MEK1/2-dependent pathway, suggesting that mechanical stress contributes to enhanced YKL-40 levels in asthmatic lungs [65]. A further mechanistic study found that YKL-40 increased the proliferation and migration of bronchial smooth muscle (BSM) cells through PAR-2-, AKT-, ERK-, and p38-dependent mechanisms and demonstrated that YKL-40 epithelial expression was positively correlated with BSM mass in asthma[66]. Other studies showed that YKL-40 induced IL-8/CXCL8 expression from bronchial epithelium via MAPK (JNK and ERK) and NF- B pathways [67]. Taken together, these results suggest that YKL-40-mediated IL-8 production could be related to BSM remodeling [68]. Further studies showed that the allergen ovalbumin increased YKL-40 expression in tracheal epithelial cells [69] and demonstrated that YKL-40 increased mucin5AC production in human bronchial epithelial cells [70]. Collectively, these studies extend the view that YKL-40 is mainly a marker of neutrophilic inflammation by demonstrating modulatory effects of this chitinase-like protein on airway epithelial cells. Despite these intriguing insights into the biological effects of chitinase-like proteins, their precise functional role in biological processes and disease conditions in humans still remains largely unclear.

CONCLUSIONS

Chitin, chitinases, and chitinase-like proteins remain enigmatic terms for human diseases. However, after the second look into the pathophysiology of allergic and chronic lung diseases, these ancient, insect glycoprotein-associated pathways attract high relevance as potential biomarkers and therapeutic targets. Particularly, fungal chitin-linked asthma is increasing, but treatment options and successful clinical trials are scarce, necessitating further therapeutic developments. Before those approaches can be applied clinically, several key questions remain to be answered:

How is chitin recognized by the human immune system? Is it a novel pattern recognition receptor ligand? And if so, can this interaction be targeted and exploited therapeutically?

What is the primary role of enzymatically active chitinase found in the human body? To defend against chitin-bearing pathogens or to skew the immune system?

Is a dysregulation of chitin sensing or YKL-40 induction pathways associated with altered susceptibility for diseases like fungal asthma or house-dust mite allergy?

Is YKL-40 a potential biomarker of severe asthma, a marker of neutrophilic innate immune activation, or a marker reflecting tissue remodeling (or all of these)?

Does YKL-40 play a causative role in these disorders? Can YKL-40 be neutralized pharmacologically? And if so, which diseases would benefit?

Considering YKL-40 as a therapeutic target in human diseases, what is the physiologic function of this protein?

These issues remain to be solved and pave the way for an exciting novel field in pediatric immunology, bridging a gap between insects, fungi, immune cells, and the lung.

AUTHORS' CONTRIBUTIONS

IM and DH wrote the major parts of the manuscript and coordinated and supervised the writing process of the whole manuscript. AH, MB, JK, KF, ANW, and MM contributed to specific sections of the review. All authors read and approved the final manuscript.

ACKNOWLEDGEMENTS

We thank Peter Weber, Department of Pediatrics, Tübingen, Germany, for creating and illustrating the figure. We thank the SFB/CRC685 (D.H.), University of Tübingen, for support.

REFERENCES

1. Lee CG, Da Silva CA, Lee JY, Hartl D, Elias JA (2008) Chitin regulation of immune responses: an old molecule with new roles. Curr Opin Immunol 20(6):684-689b doi:10.1016/j.coi.2008.10.002
2. Lee CG (2009) Chitin, chitinases and chitinase-like proteins in allergic inflammation and tissue remodeling. Yonsei Med J 50(1):22-30 doi:10.3349/ymj.2009.50.1.22
3. Denning DW, Pashley C, Hartl D, Wardlaw A, Godet C, Del Giacco S, Delhaes L, Sergejeva S (2014) Fungal allergy in asthma-state of the art and research needs. Clin Transl Allergy 4:14 doi:10.1186/2045-7022-4-14

4. Shibata Y, Foster LA, Metzger WJ, Myrvik QN (1997) Alveolar macrophage priming by intravenous administration of chitin particles, polymers of *N*-acetyl-D-glucosamine, in mice. Infect Immun 65(5):1734-1741

5. Reese TA, Liang HE, Tager AM, Luster AD, Van Rooijen N, Voehringer D, Locksley RM (2007) Chitin induces accumulation in tissue of innate immune cells associated with allergy. Nature 447(7140):92-96

6. Van Dyken SJ, Garcia D, Porter P, Huang X, Quinlan PJ, Blanc PD, Corry DB, Locksley RM (2011) Fungal chitin from asthma-associated home environments induces eosinophilic lung infiltration. J Immunol 187(5):2261-2267 doi:10.4049/jimmunol.1100972

7. Shibata Y, Foster LA, Bradfield JF, Myrvik QN (2000) Oral administration of chitin down-regulates serum IgE levels and lung eosinophilia in the allergic mouse. J Immunol 164(3):1314-1321

8. Strong P, Clark H, Reid K (2002) Intranasal application of chitin microparticles down-regulates symptoms of allergic hypersensitivity to *Dermatophagoides pteronyssinus* and*Aspergillus fumigatus* in murine models of allergy. Clin Exp Allergy 32(12):1794-1800

9. Ozdemir C, Yazi D, Aydogan M, Akkoc T, Bahceciler NN, Strong P, Barlan IB (2006) Treatment with chitin microparticles is protective against lung histopathology in a murine asthma model. Clin Exp Allergy 36(7):960-968 doi:10.1111/j.1365-2222.2006.02515.x

10. Wan J, Zhang XC, Neece D, Ramonell KM, Clough S, Kim SY, Stacey MG, Stacey G (2008) A LysM receptor-like kinase plays a critical role in chitin signaling and fungal resistance in*Arabidopsis*. Plant Cell 20(2):471-481 doi:10.1105/tpc.107.056754

11. Miya A, Albert P, Shinya T, Desaki Y, Ichimura K, Shirasu K, Narusaka Y, Kawakami N, Kaku H, Shibuya N (2007) CERK1, a LysM receptor kinase, is essential for chitin elicitor signaling in *Arabidopsis*. Proc Natl Acad Sci U S A 104(49):19613-19618 doi:10.1073/pnas.0705147104

12. Wagener J, Malireddi RK, Lenardon MD, Koberle M, Vautier S, MacCallum DM, Biedermann T, Schaller M, Netea MG, Kanneganti TD, Brown GD, Brown AJ, Gow NA (2014) Fungal chitin dampens inflammation through IL-10 induction mediated

by NOD2 and TLR9 activation. Plos Pathog 10(4):e1004050 doi:10.1371/journal.ppat.1004050

13. Zheng T, Rabach M, Chen NY, Rabach L, Hu X, Elias JA, Zhu Z (2005) Molecular cloning and functional characterization of mouse chitotriosidase. Gene 357(1):37-46
14. Herrera-Estrella A, Chet I (1999) Chitinases in biological control. EXS 87:171-184
15. Boot RG, Renkema GH, Strijland A, van Zonneveld AJ, Aerts JM (1995) Cloning of a cDNA encoding chitotriosidase, a human chitinase produced by macrophages. J Biol Chem 270(44):26252-26256
16. Boot RG, Blommaart EF, Swart E, Ghauharali-van der Vlugt K, Bijl N, Moe C, Place A, Aerts JM (2001) Identification of a novel acidic mammalian chitinase distinct from chitotriosidase. J Biol Chem 276(9):6770-6778
17. Bleau G, Massicotte F, Merlen Y, Boisvert C (1999) Mammalian chitinase-like proteins. EXS 87:211-221
18. van Eijk M, van Roomen CP, Renkema GH, Bussink AP, Andrews L, Blommaart EF, Sugar A, Verhoeven AJ, Boot RG, Aerts JM (2005) Characterization of human phagocyte-derived chitotriosidase, a component of innate immunity. Int Immunol 17(11):1505-1512 doi:10.1093/intimm/dxh328
19. van Eijk M, Scheij SS, van Roomen CP, Speijer D, Boot RG, Aerts JM (2007) TLR- and NOD2-dependent regulation of human phagocyte-specific chitotriosidase. Febs Lett 581(28):5389-5395 doi:10.1016/j.febslet.2007.10.039
20. Volck B, Price PA, Johansen JS, Sorensen O, Benfield TL, Nielsen HJ, Calafat J, Borregaard N (1998) YKL-40, a mammalian member of the chitinase family, is a matrix protein of specific granules in human neutrophils. Proc Assoc Am Physicians 110(4):351-360
21. Lee CG, Hartl D, Lee GR, Koller B, Matsuura H, Da Silva CA, Sohn MH, Cohn L, Homer RJ, Kozhich AA, Humbles A, Kearley J, Coyle A, Chupp G, Reed J, Flavell RA, Elias JA (2009) Role of breast regression protein 39 (BRP-39)/chitinase 3-like-1 in Th2 and IL-13-induced tissue responses and apoptosis. J Exp Med 206(5):1149-1166 doi:jem.20081271

22. Hartl D, Lee CG, Da Silva CA, Chupp GL, Elias JA (2009) Novel biomarkers in asthma: chemokines and chitinase-like proteins. Curr Opin Allergy Clin Immunol 9(1):60-66 doi:10.1097/ACI.0b013e32831f8ee0

23. Zhu Z, Zheng T, Homer RJ, Kim YK, Chen NY, Cohn L, Hamid Q, Elias JA (2004) Acidic mammalian chitinase in asthmatic Th2 inflammation and IL-13 pathway activation. Science 304(5677):1678-1682

24. Lee CG, Da Silva CA, Dela Cruz CS, Ahangari F, Ma B, Kang MJ, He CH, Takyar S, Elias JA (2011) Role of chitin and chitinase/chitinase-like proteins in inflammation, tissue remodeling, and injury. Annu Rev Physiol 73:479-501 doi:10.1146/annurev-physiol-012110-142250

25. Chupp GL, Lee CG, Jarjour N, Shim YM, Holm CT, He S, Dziura JD, Reed J, Coyle AJ, Kiener P, Cullen M, Grandsaigne M, Dombret MC, Aubier M, Pretolani M, Elias JA (2007) A chitinase-like protein in the lung and circulation of patients with severe asthma. N Engl J Med 357(20):2016-2027

26. Konradsen JR, James A, Nordlund B, Reinius LE, Soderhall C, Melen E, Wheelock AM, Lodrup Carlsen KC, Lidegran M, Verhoek M, Boot RG, Dahlen B, Dahlen SE, Hedlin G (2013) The chitinase-like protein YKL-40: a possible biomarker of inflammation and airway remodeling in severe pediatric asthma. J Allergy Clin Immunol 132(2):328-335 e325. doi:10.1016/j.jaci.2013.03.003

27. Santos CB, Davidson J, Covar RA, Spahn JD (2014) The chitinase-like protein YKL-40 is not a useful biomarker for severe persistent asthma in children. Ann Allergy Asthma Immunol 113(3):263-266 doi:10.1016/j.anai.2014.05.024

28. Goldman DL, Li X, Tsirilakis K, Andrade C, Casadevall A, Vicencio AG (2012) Increased chitinase expression and fungal-specific antibodies in the bronchoalveolar lavage fluid of asthmatic children. Clin Exp Allergy 42(4):523-530 doi:10.1111/j.1365-2222.2011.03886.x

29. Kuepper M, Bratke K, Virchow JC (2008) Chitinase-like protein and asthma. N Engl J Med 358(10):1073-1075 author reply 1075. doi:10.1056/NEJMc073406

30. Gavala ML, Kelly EA, Esnault S, Kukreja S, Evans MD, Bertics PJ, Chupp GL, Jarjour NN (2013) Segmental allergen challenge

enhances chitinase activity and levels of CCL18 in mild atopic asthma. Clin Exp Allergy 43(2):187-197 doi:10.1111/cea.12032

31. Ober C, Tan Z, Sun Y, Possick JD, Pan L, Nicolae R, Radford S, Parry RR, Heinzmann A, Deichmann KA, Lester LA, Gern JE, Lemanske RF Jr, Nicolae DL, Elias JA, Chupp GL (2008) Effect of variation in CHI3L1 on serum YKL-40 level, risk of asthma, and lung function. N Engl J Med 358(16):1682-1691
32. Tsai Y, Ko Y, Huang M, Lin M, Wu C, Wang C, Chen Y, Li J, Tseng Y, Wang T (2014) CHI3L1 polymorphisms associate with asthma in a Taiwanese population. BMC Med Genet 15:86 doi:10.1186/1471-2350-15-86
33. Rathcke CN, Holmkvist J, Husmoen LL, Hansen T, Pedersen O, Vestergaard H, Linneberg A (2009) Association of polymorphisms of the CHI3L1 gene with asthma and atopy: a populations-based study of 6514 Danish adults. Plos One 4(7):e6106 doi:10.1371/journal.pone.0006106
34. Tang H, Fang Z, Sun Y, Li B, Shi Z, Chen J, Zhang T, Xiu Q (2010) YKL-40 in asthmatic patients, and its correlations with exacerbation, eosinophils and immunoglobulin E. Eur Respir J 35(4):757-760 doi:10.1183/09031936.00034409
35. Specjalski K, Jassem E (2011) YKL-40 protein is a marker of asthma. J Asthma 48(8):767-772 doi:10.3109/02770903.2011.611955
36. Matsuura H, Hartl D, Kang MJ, Dela Cruz CS, Koller B, Chupp GL, Homer RJ, Zhou Y, Cho WK, Elias JA, Lee CG (2011) Role of breast regression protein-39 in the pathogenesis of cigarette smoke-induced inflammation and emphysema. Am J Resp Cell Mol 44(6):777-786 doi:10.1165/rcmb.2010-0081OC
37. Letuve S, Kozhich A, Arouche N, Grandsaigne M, Reed J, Dombret MC, Kiener PA, Aubier M, Coyle AJ, Pretolani M (2008) YKL-40 is elevated in patients with chronic obstructive pulmonary disease and activates alveolar macrophages. J Immunol 181(7):5167-5173 doi:181/7/5167
38. Furuhashi K, Suda T, Nakamura Y, Inui N, Hashimoto D, Miwa S, Hayakawa H, Kusagaya H, Nakano Y, Nakamura H, Chida K (2010) Increased expression of YKL-40, a chitinase-like protein, in serum and lung of patients with idiopathic pulmonary fibrosis. Respir Med 104(8):1204-1210 doi:10.1016/j.rmed.2010.02.026

39. Kim HR, Jun CD, Lee KS, Cho JH, Jeong ET, Yang SH, Lee YJ, Park DS (2012) Levels of YKL-40 in pleural effusions and blood from patients with pulmonary or pleural disease. Cytokine 58(3):336-343 doi:10.1016/j.cyto.2012.03.001

40. Xu CH, Yu LK, Hao KK (2014) Serum YKL-40 level is associated with the chemotherapy response and prognosis of patients with small cell lung cancer. Plos One 9(5):e96384 doi:10.1371/journal.pone.0096384

41. Thom I, Andritzky B, Schuch G, Burkholder I, Edler L, Johansen JS, Bokemeyer C, Schumacher U, Laack E (2010) Elevated pretreatment serum concentration of YKL-40-An independent prognostic biomarker for poor survival in patients with metastatic nonsmall cell lung cancer. Cancer 116(17):4114-4121 doi:10.1002/cncr.25196

42. Jaksch P, Taghavi S, Klepetko W, Salama M (2014) Pretransplant serum human chitinase-like glycoprotein YKL-40 concentrations independently predict bronchiolitis obliterans development in lung transplant recipients. J Thorac Cardiovasc Surg 148(1):273-281 doi:10.1016/j.jtcvs.2014.02.059

43. Sohn MH, Kang MJ, Matsuura H, Bhandari V, Chen NY, Lee CG, Elias JA (2010) The chitinase-like proteins breast regression protein-39 and YKL-40 regulate hyperoxia-induced acute lung injury. Am J Respir Crit Care Med 182(7):918-928 doi:10.1164/rccm.200912-1793OC

44. Kruit A, Grutters JC, Ruven HJ, van Moorsel CC, van den Bosch JM (2007) A CHI3L1 gene polymorphism is associated with serum levels of YKL-40, a novel sarcoidosis marker. Respir Med 101(7):1563-1571 doi:10.1016/j.rmed.2006.12.006

45. Park SJ, Jun YJ, Kim TH, Jung JY, Hwang GH, Jung KJ, Lee SH, Lee HM, Lee SH (2013) Increased expression of YKL-40 in mild and moderate/severe persistent allergic rhinitis and its possible contribution to remodeling of nasal mucosa. Am J Rhinol Allergy 27(5):372-380 doi:10.2500/ajra.2013.27.3941

46. Hector A, Kormann MSD, Mack I, Latzin P, Casaulta C, Kieninger E, Zhou Z, Yildirim AO, Bohla A, Rieber N, Kappler M, Koller B, Eber E, Eickmeier O, Zielen S, Eickelberg O, Griese M, Mall MA, Hartl D (2011) The Chitinase-Like Protein YKL-40 Modulates Cystic Fibrosis Lung Disease. Plos One 6 (9). doi:ARTN e24399 DOI 10.1371/journal.pone.0024399

47. Elias JA, Homer RJ, Hamid Q, Lee CG (2005) Chitinases and chitinase-like proteins in T(H)2 inflammation and asthma. J Allergy Clin Immunol 116(3):497-500 doi:10.1016/j.jaci.2005.06.028
48. Mall MA, Hartl D (2014) CFTR: cystic fibrosis and beyond. Eur Respir J 44(4):1042-1054 doi:10.1183/09031936.00228013
49. Hartl D, Gaggar A, Bruscia E, Hector A, Marcos V, Jung A, Greene C, McElvaney G, Mall M, Doring G (2012) Innate immunity in cystic fibrosis lung disease. J Cyst Fibros 11(5):363-382 doi:10.1016/j.jcf.2012.07.003
50. Fantino E, Gangell CL, Hartl D, Sly PD, Arest CF (2014) Airway, but not serum or urinary, levels of YKL-40 reflect inflammation in early cystic fibrosis lung disease. BMC Pulm Med 14:28 doi:10.1186/1471-2466-14-28
51. Zhou Z, Duerr J, Johannesson B, Schubert SC, Treis D, Harm M, Graeber SY, Dalpke A, Schultz C, Mall MA (2011) The ENaC-overexpressing mouse as a model of cystic fibrosis lung disease. J Cyst Fibros 10(Suppl 2):S172-S182 doi:10.1016/S1569-1993(11)60021-0
52. Mall M, Grubb BR, Harkema JR, O'Neal WK, Boucher RC (2004) Increased airway epithelial Na+ absorption produces cystic fibrosis-like lung disease in mice. Nat Med 10(5):487-493 doi:10.1038/nm1028nm1028
53. Mall MA, Graeber SY, Stahl M, Zhou-Suckow Z (2014) Early cystic fibrosis lung disease: role of airway surface dehydration and lessons from preventive rehydration therapies in mice. Int J Biochem Cell Biol 52:174-179 doi:10.1016/j.biocel.2014.02.006
54. Mall MA, Harkema JR, Trojanek JB, Treis D, Livraghi A, Schubert S, Zhou Z, Kreda SM, Tilley SL, Hudson EJ, O'Neal WK, Boucher RC (2008) Development of chronic bronchitis and emphysema in beta-epithelial Na+ channel-overexpressing mice. Am J Respir Crit Care Med 177(7):730-742 doi:200708-1233OC
55. Mall MA (2008) Role of cilia, mucus, and airway surface liquid in mucociliary dysfunction: lessons from mouse models. J Aerosol Med Pulm D 21(1):13-24 doi:10.1089/jamp.2007.0659
56. Livraghi-Butrico A, Kelly EJ, Klem ER, Dang H, Wolfgang MC, Boucher RC, Randell SH, O'Neal WK (2012) Mucus clearance, MyD88-dependent and MyD88-independent immunity

modulate lung susceptibility to spontaneous bacterial infection and inflammation. Mucosal Immunol 5(4):397-408 doi:10.1038/mi.2012.17

57. Trojanek JB, Cobos-Correa A, Diemer S, Kormann M, Schubert SC, Zhou-Suckow Z, Agrawal R, Duerr J, Wagner CJ, Schatterny J, Hirtz S, Sommerburg O, Hartl D, Schultz C, Mall MA (2014) Airway mucus obstruction triggers macrophage activation and MMP12-dependent emphysema. Am J Respir Cell Mol Biol. doi:10.1165/rcmb.2013-0407OC

58. Saini Y, Dang H, Livraghi-Butrico A, Kelly EJ, Jones LC WKON, Boucher RC (2014) Gene expression in whole lung and pulmonary macrophages reflects the dynamic pathology associated with airway surface dehydration. BMC Genomics 15(1):726 doi:10.1186/1471-2164-15-726

59. Samstad EO, Niyonzima N, Nymo S, Aune MH, Ryan L, Bakke SS, Lappegard KT, Brekke OL, Lambris JD, Damas JK, Latz E, Mollnes TE, Espevik T (2014) Cholesterol crystals induce complement-dependent inflammasome activation and cytokine release. J Immunol 192(6):2837-2845 doi:10.4049/jimmunol.1302484

60. De Nardo D, De Nardo CM, Latz E (2014) New insights into mechanisms controlling the NLRP3 inflammasome and its role in lung disease. Am J Pathol 184(1):42-54 doi:10.1016/j.ajpath.2013.09.007

61. Gross O, Thomas CJ, Guarda G, Tschopp J (2011) The inflammasome: an integrated view. Immunol Rev 243(1):136-151 doi:10.1111/j.1600-065X.2011.01046.x

62. Guerra S, Halonen M, Sherrill DL, Venker C, Spangenberg A, Carsin AE, Tares L, Lavi I, Barreiro E, Martinez-Moratalla J, Urrutia I, Sunyer J, Anto JM, Martinez FD (2013) The relation of circulating YKL-40 to levels and decline of lung function in adult life. Respir Med 107(12):1923-1930 doi:10.1016/j.rmed.2013.07.013

63. Cho SJ, Nolan A, Echevarria GC, Kwon S, Naveed B, Schenck E, Tsukiji J, Prezant DJ, Rom WN, Weiden MD (2013) Chitotriosidase is a biomarker for the resistance to World Trade Center lung injury in New York City firefighters. J Clin Immunol 33(6):1134-1142 doi:10.1007/s10875-013-9913-2

64. Wu AC, Lasky-Su J, Rogers CA, Klanderman BJ, Litonjua AA (2010) Fungal exposure modulates the effect of polymorphisms of

chitinases on emergency department visits and hospitalizations. Am J Respir Crit Care Med 182(7):884-889 doi:10.1164/rccm.201003-0322OC

65. Park JA, Drazen JM, Tschumperlin DJ (2010) The chitinase-like protein YKL-40 is secreted by airway epithelial cells at base line and in response to compressive mechanical stress. J Biol Chem 285(39):29817-29825 doi:10.1074/jbc.M110.103416

66. Bara I, Ozier A, Girodet PO, Carvalho G, Cattiaux J, Begueret H, Thumerel M, Ousova O, Kolbeck R, Coyle AJ, Woods J, Tunon de Lara JM, Marthan R, Berger P (2012) Role of YKL-40 in bronchial smooth muscle remodeling in asthma. Am J Respir Crit Care Med 185(7):715-722 doi:10.1164/rccm.201105-0915OC

67. Tang H, Sun Y, Shi Z, Huang H, Fang Z, Chen J, Xiu Q, Li B (2013) YKL-40 induces IL-8 expression from bronchial epithelium via MAPK (JNK and ERK) and NF-kappaB pathways, causing bronchial smooth muscle proliferation and migration. J Immunol 190(1):438-446 doi:10.4049/jimmunol.1201827

68. Tang H, Shi Z, Xiu Q, Li B, Sun Y (2012) YKL-40-mediated interleukin 8 production may be closely associated with remodeling of bronchial smooth muscle cells. Am J Respir Crit Care Med 186(4):386 author reply 386–387. doi:10.1164/ajrccm.186.4.386a

69. Ben SQ, Qiu YL, Zhou J, Zhou XY, Zhang S, Wu Y, Ju SQ, Ni SS (2014) Ovalbumin enhances YKL-40, IL-5, GM-CSF, and eotaxin expression simultaneously in primarily cultured mouse tracheal epithelial cells. In Vitro Cell Dev Biol Anim 50(3):243-250 doi:10.1007/s11626-013-9698-x

70. Liu C, Li Q, Zhou X, Kolosov VP, Perelman JM (2013) The chitinase-like protein YKL-40 increases mucin5AC production in human bronchial epithelial cells. Exp Cell Res 319(18):2866-2873 doi:10.1016/j.yexcr.2013.08.009.

Chapter 7

Digestion of Bio-waste - GHG Emissions and Mitigation Potential

Jaqueline Daniel-Gromke, Jan Liebetrau, Velina Denysenko, and Christian Krebs

Department Biochemical Conversion, DBFZ Deutsches Biomasseforschungszentrum gGmbH, Torgauer Straße 116, Leipzig, 04347, Germany

ABSTRACT

Background

For a precise description of the emission situation of the anaerobic digestion (AD) of the separately collected organic fraction of household waste (bio-waste), only a few data are available. The paper presents the greenhouse gas (GHG) emissions measured at 12 representative AD plants treating bio-waste. The results of the emission measurements were used to assess the ecological impact of bio-waste digestion and to describe possible mitigation measures to reduce the occurring GHG emissions. With respect to the climate protection, a quantitative assessment of the emissions of energy generation from biomass and biological waste treatment is important. Biogas plants need to be operated in a way that negative environmental effects are avoided and human health is not compromised.

Methods

GHG balances were calculated based on the measured emissions of the gases methane, nitrous oxide, and ammonia of bio-waste AD plants. The emission analysis supports the reduction of GHGs in biogas production and contributes to a climate-efficient technology.

Results

The results show that GHG emissions can be minimized, if the technology and operation of the plant are adjusted accordingly. The open storage of active material (e.g., insufficient fermented residues from batch fermentation systems), open digestate storage tanks, missing acidic scrubbers in front of bio-filters, or insufficient air supply during the post-composting of digestate can cause relevant GHG emissions.

Conclusions

Consequently avoiding open storage of insufficient fermented residues and using aerated post-composting with short turnover periods, smaller heaps, and an optimized amount of structure (woody) material can reduce GHG emissions.

BACKGROUND

Gaseous emissions are of great importance referring to the operation of biogas plants because they can affect the safety, the greenhouse gas

(GHG) balance, and the economy of plants significantly. Depending on the used technology and the kind of operation, GHG emissions like methane, nitrous oxide, and ammonia are occurring. Methane emissions dominate GHG emissions of biogas plants.

Due to the global warming potential (GWP) of 25 relative to carbon dioxide [1], methane emissions have a strong effect on the climate change. Leakages, process disturbances, and unavoidable emissions during operation can influence the total GHG performance of the biogas plant negatively. Regarding measured emissions of biogas plants in operation, only a small number of detailed studies are available.

In former studies, the overall emissions of biogas plants usually have been estimated by assumptions, e.g., '1 % of diffuse methane emissions from the components of anaerobic digestion (AD) plants like digester, pipes,' etc. (e.g. [2],[3]). However, in the recent years, several studies estimated methane emissions from biogas plants (e.g. [4]-[9]). Most of the published studies analyzed agricultural AD plants; if waste treating plants were investigated, only a few AD components were monitored as summarized by Dumont et al. [10]. Due to the fact that there are only few data describing the emission situation of AD plants based on bio-waste, in the study described here, 12 representative bio-waste treatment plants with AD process as part of the overall operation were analyzed. The overall objective of the study was a detailed analysis of GHG emissions generated from biogas production from bio-waste. This paper presents the results of a comprehensive measurement for GHG emissions at bio-waste digestion plants in operation during a long-term period of 3 years. Representative bio-waste digestion plants have been selected, and all relevant components of the process chain were investigated during two periods of a week per year on each of the selected plants to identify the main emission sources and the quantity of the emissions. The results of the emission measurements were implemented in an ecological assessment focused on GHG balances. The results of the examined biogas plants allow a description of possible mitigation measures to reduce GHG emissions. The results bring new aspects into the actual data base to support the assessment of environmental impacts of bio-waste digestion. Thus, the tests on practice biogas plants with respect to the whole process chain allow an optimization of the process in terms of reducing any identified emissions.

In Germany, approximately 9 million tons of bio-waste and green waste per year were collected separately in 2011 [11]. Most of this collected bio-waste and green cuts are used in composting processes. About 1.15 million tons of bio-waste per year and 0.05 tons of green cuts per year are used for digestion in biogas facilities [12]. By the end of 2013, there have been about 130 plants generating biogas from organic waste in operation. Compared to agricultural biogas plants, there is a higher share of dry fermentation processes in AD plants based on bio-waste. About one half of the bio-waste digestion plants are operated as dry fermentation plants in Germany, whereas half of the dry fermentation plants are operated discontinuously (batch system). Currently, there are 25 batch systems based on bio-waste in operation [13]. Due to the robustness of the process and the possibility to treat substrates which are hardly pumpable and contain disturbing materials (e.g., stones, metals, glass), the use of batch systems in case of dry fermentation processes of bio-waste is increasing. In the future, it will be more important to exploit additional potentials in the field of organic waste and residues from industry and municipalities. In the field of municipal bio-waste, the exploitation of additional potentials is in progress. The amount of municipal bio-waste that is available for digestion in biogas plants will increase considerably within the next years. Currently, a considerable trend to digestion of bio-waste and green waste, often integrated as so-called upstream systems into existing composting plants, can be assessed.

METHODS

Twelve biogas plants were selected for the detection of plant-based emissions of methane (CH_4), nitrous oxide (N_2O), and ammonia (NH_3). Based on the measured emission rates, GHG balances in compliance with the analysis of GHG credits (e.g., for biogas production, fertilizer, and humus effect of fermentation products and composts) were prepared. Thus, the electricity production and heat utilization of biogas as well as the credits of the various fermentation residues were analyzed to estimate the specific GHG performance of the investigated facilities. Finally, the measurements with respect to mitigation of GHG emissions were analyzed and described.

Investigated Biogas Plants

The emission analysis includes four continuously operated wet fermentation plants (continuous stirred-tank reactor, CSTR), five continuous dry fermentation plants (plug-flow fermenter), and three batch fermentation processes (discontinuous operation, 'garage style' digesters). Table 1 shows the investigated 12 AD plants based on bio-waste with their specific characteristics. Table 2 presents the amount and kind of substrate treated at the bio-waste facility. The treated bio-waste is used completely for digestion in AD plant nos. 2, 4, and 5. Most AD plants operate with partial stream digestion of bio-waste. In these plants, just the bio-waste from separate collection is used for fermentation, whereas the green cut and structure (woody) material is added after digestion within the composting process.

Table 1: Characteristics of investigated AD plants based on bio-waste

Plant no.	Installed electrical capacity kWel	Kind of fermentationa	Temperatureb	Range of temp. in °Cc	Mode of operation	HRT in daysd	**Residues storage tank**	**Post-compostinge**	**Type of aeration (post-composting)**	**External heat utilizationf**
1	630	Wet	M		Multi-stage	8	Covered	x	Open, unaerated	
2	536	Wet	T		Multi-stage	20	Covered	x	Open, unaerated	x
3	986	Wet	M		Single-stage	17	Open			
4	1200	Wet	M	37-40	Multi-stage	25	Open, covered			
5	1790	Dry	T		Single-stage	25	Gas-proof covered			
6	1413	Dry	T	55	Single-stage	21	Covered	x	Open, unaerated	
7	816	Dry	T		Single-stage	28		x	Enclosed, aerated (pressure ventilation)	
8	625	Dry	T	55	Single-stage	14	Gas-proof covered	x	Enclosed, aerated and unaerated (pressure ventilation)	x

9	640	Dry	T		Single-stage	21	Covered	x	Enclosed, aerated (pressure ventilation)	
10	625	Batch	M	37-39	Single-stage	28		x	Enclosed, aerated (pressure ventilation)	x
11	680	Batch	M	40-42	Single-stage	21		x	Open, aerated, enclosed	x
12	370	Batch	M	40-42	Single-stage	21		x	Open, unaerated	x

[a]Wet = wet fermentation, dry = dry fermentation, batch = batch system (discontinuous). [b]M = mesophilic, T = thermophilic. [c]According to the information of plant operators (if available). [d]Hydraulic retention time. [e]x = post-composting process. [f]x = external heat utilization.

Daniel-Gromke *et al.*

Daniel-Gromke et al. Energy, Sustainability and Society 2015 5:3, doi: 10.1186/s13705-014-0032-6

Table 2: Amount and kind of treated substrate of investigated bio-waste facilities

Plant no.	Total amount of substrate input treated at facility t/a (fresh matter)	Amount of input for AD t/a (fresh matter)	Percent share of AD (mass) related to total amount treated at facility	Kind of substratea
1	83.840	32.000	38	Bio-waste, green cut
2	10.062	9.865	98	Bio-waste, catering waste, green cut
3	34.976	25.702	73	Bio-waste
4	35.388	35.388	100	Sludge from wastewater treatment, catering waste, food waste, manure
5	33.130	33.130	100	Bio-waste, catering waste
6	29.900	26.910	90	Bio-waste, green cut
7	35.450	17.725	50	Bio-waste
8	20.000	12.000	60	Bio-waste, green cut
9	23.000	17.250	75	Bio-waste, green cut
10	36.000	18.000	50	Bio-waste, green cut
11	73.333	22.000	30	Bio-waste, green cut
12	13.333	12.000	90	Bio-waste, green cut, food-waste

[a]Bio-waste = bio-waste from separate collection.

Daniel-Gromke et al.

Daniel-Gromke et al. Energy, Sustainability and Society 2015 5:3, doi: 10.1186/s13705-014-0032-6

AD plant nos. 1, 2, and 12 were operated with open, unaerated post-composting processes. AD plant no. 3 had a covered but no enclosed composting steps. In AD plant no. 4, larger quantities of sludge from wastewater treatment were treated. Thus, primarily liquid digestate was generated. The small amounts of solid digestate were stored on site and were used for external composting. The solid digestate of AD plant no. 5 were stored open after separation. Post-composting processes with active ventilation (pressure ventilation) and enclosed composting systems were used at AD plant nos. 7, 9, and 10. A defined step of aeration in which the air is integrated into the exhaust gas treatment (bio-filter) was considered at plant no. 10.

All investigated biogas facilities operated with bio-filters as gas treatment. However, most of plant operators did not use acidic scrubbers at biogas facilities. Only four of 12 plants operated with acidic scrubbers, and the proper operation was not always ensured. Five plants used the bio-filter combined with humidifier. The exhaust gas should be treated with acid scrubbers to deposit NH_3 and minimize N_2O formation in the bio-filter (e.g., plant nos. 5 and 9). It should be recognized that there were also diffuse emission sources which were not collected by bio-filters (e.g., open doors of delivery hall at AD plant nos. 6 and 7; post-composting at AD plant nos. 8, 9, 11).

Often, digestate - whether separated or not separated - is stored open temporarily or for longer periods. Four of the seven examined plants which stored liquid digestate or process waters used covered storage tank (AD plant nos. 4, 5, 8, and 9). Two plants (nos. 5 and 8) with gas-proof covered storage tank are able to use the exhaust gas by involving into the CHP.

Emission Measurements

There are in general two methods to determine the emissions of a large industrial facility or areas with diffuse emission sources. One way is to attempt to capture the overall emissions of the facilities by means of concentration measurements in the surroundings and the application of inverse dispersion models [7] or radial plume mapping [14]. These methods allow the determination of the overall emissions of a large area with uncertain sources of emission. They do not allow the localization of single sources and allocation of a certain quantity to them. However, for further efficient measures to reduce emissions, it is very important to identify and quantify the emission sources on site. For this reason, the methods used focus on the identification and quantification of single sources [5].

The emission analysis included two measurement periods in each plant (each 1 week in 2010 and 2011), in which all plant components from substrate delivery to storage of digestate and composting were investigated. The measured emissions of both periods were averaged. Several sampling points at AD plant and compost heaps were examined. Following the inspection of the biogas facilities on site, potential significant emission sources within the process chain were identified. The following emission sources were investigated: delivery and conditioning of substrate (material handling), storage of fermentation residues (digestate), fermenter, before and after exhaust gas treatment (acid scrubber and bio-filter), and exhaust of CHP unit (combined heat and power plant) as well as post-composting process of digestate. The emission measurements focused on the emission detection at the AD plant and post-composting processes - not the utilization of biogas in CHP units. Therefore, not all CHP were measured. With respect to the total GHG balance, the production as well as the utilization of biogas in CHP is important. Thus, an average of CHP emissions was considered (see 'Emissions from CHP'). For the emission measurements of the composting process, four or five sections of the windrow were selected for each measurement period, which differed in time of composting resp. age of rotting material.

According to the characteristics of the gases, the applied measurement techniques were adjusted. Leakage detection techniques were used to find the critical spots within the process; open and closed domes were used to determine the main emission sources. Regarding the methods of emission measurements, there are differences between captured and diffuse emission sources. Accordingly, different measurements for emissions from encapsulated areas (e.g., delivery hall with collection of exhaust) and diffuse emission sources during several measured periods were used. Waste treatment facilities often have gas collection systems that collect air from the captured process steps and deliver the gas after a cleaning stage into the atmosphere. In most cases, the cleaning step is a bio-filter. Because of that, in all investigated AD plants, the exhaust streams before and after treatment by bio-filters were examined. Depending on the plant system, further sampling points were analyzed.

In case of encapsulated emission sources, the exhaust air flow was examined directly. Thereby, the volume flow and mass concentration within the investigated pipelines were determined. The volume flows were measured with vane anemometers. The quantity of the emission source was calculated from the concentration difference and the flow rate of the blower by using the following equation [5].

$$F = Q * \rho * (c_{out} - c_{in}) \quad (1)$$

F, emission flow rate (mg/h); *Q*, air flow rate (m^3/h); *ρ*, density of the target gas (kg/m^3); c_{out} exhaust gas concentration (mg/kg); c_{in}, background gas concentration (mg/kg).

Emissions of post-composting with active aeration (e.g., actively ventilated tunnel or container systems) were measured by using encapsulated areas with air extraction. In case of open windrows composting without active aeration, a wind tunnel as emission measurement was used. An air flow was generated by using a ventilator. The measurement methods, techniques, and technical guidelines used for the determination of emission concentrations are shown in Table 3. CH_4 was detected by gas chromatography with a flame ionization detector (FID), N_2O by gas chromatography, and NH_3 by absorption in an acid solution. The sampling for the determination of CH_4 and N_2O was carried out by a measuring gas line which is connected to a gas analysis with online data collection. The sampling for the determination of NH_3 occurs directly at the tunnel exit. The sample gas is guided without gas cooling through two wash bottles filled with sulfuric acid. Further information according to the methods of emission measurement at biogas plants are published in [4].

Table 3: Measurement methods, techniques and technical guidelines for the determination of emissions at the investigated AD plants [15]

Compound	Kind of determination	Measurement methods	Measurement techniques	German technical guideline used for the calculation of emissions
Total carbon	Continuously, online data	Flame ionization detection (FID)	Bernath Atomic 3006	VDI 3481 - 3, VDI 3481 - 4, DIN EN 12619, DIN EN 13526

Methane	Continuously, online data	Infra-red (IR) method	ABB Advance Optima URAS 14	
Nitrous oxide	Continuously, online data	Infra-red (IR) spectroscopy	ABB Advance Optima URAS 14	DIN EN ISO 21258
Methane	Discontinuously, laboratory analysis	Gas chromatography (GC) with autosampler	Sampling with evacuated vials	DIN EN ISO 25139
Nitrous oxide	Discontinuously, laboratory analysis	Gas chromatography (GC) with autosampler	Sampling with evacuated vials	VDI 2469 - 1
Ammonia	Discontinuously, laboratory analysis	Wet chemical methods with sulfuric acid	Sampling with Desaga-pump and 2 wash bottles	VDI 3496 - 1

Daniel-Gromke *et al.*

Daniel-Gromke et al. Energy, Sustainability and Society 2015 5:3, doi: 10.1186/s13705-014-0032-6

Residual Gas Potential

The residual gas potential of digestate from anaerobic treatment of bio-waste was considered. The gas potential can be analyzed by different temperature levels as described by [16]. The temperature of the stored digestate has a great influence on the emissions. Laboratory tests within the studies of [17] and [18] showed that depending on the temperature of the digestate during storage, the emission potential can be significantly reduced. In [5], it is shown that the average CH_4 potentials obtained at 20°C represent 39% of the CH_4 potential obtained at 39°C. According to [17], the CH_4production at a temperature of 25°C is reduced to 40–50% of the value obtained at 37°C and at 10°C, the CH_4 production goes down to even 1% [5].

In this study, the residual gas potential of digestate was determined at a temperature of 38°C. The digestate samples were taken directly after the fermentation step and - in case of separation of digestate - after separation (see AD plant nos. 1, 2, and 7). With these samples, batch experiments were carried out according to the German technical guideline VDI 4630 [19]. Finally, relative residual gas potentials with respect to the used fresh matter were determined using the following assumptions: average CH_4 yield of 74 m^3 CH_4 (STP) per metric ton fresh matter bio-waste, 10% degradation of fresh matter by the fermentation stage, and a separation ratio of 20% solid digestate to 80% liquid digestate.

Assumptions - GHG Balances

Based on a survey of plant operator, additional emission-related data (e.g., energy demand, amount, and kind of heat utilization) were collected to prepare the GHG balance of each plant. For the total GHG balances, the emissions as well as credits for the kind of products (combined heat and electricity from biogas; fertilizer and humus supply from fermentation residues) were considered. The overall GHG performance of each AD plant included in particular the following: GHG emissions according to the measured components of AD plant, calculated emissions of the electricity demand (AD plant and CHP), calculated emissions during the application of the fermentation residues, credits for the electricity production from biogas (substitution of fossil electricity supply), credits for the utilization of exhaust heat (substitution of fossil heat), and credits for the use of fermentation products (substitution of fossil fertilizer and peat, humus effects).

The considered GHG emissions for all processes of bio-waste digestion were converted into CO_2equivalents (CO_2-eq) by using characterization factors.

The following factors according to the GWP for a 100-year time period were stated: $CO_2 = 1$, $CH_4 = 25$, $N_2O = 298$ [1]. With respect to the NH_3 emissions, it is assumed that 1% of the NH_3 is converted to N_2O emissions [1].

As a functional unit of GHG balances, 'ton input bio-waste treated at facility (fresh matter)' was used. This unit included the total amount of waste treated at the facility (bio-waste and green waste - if any) - not only the amount of bio-waste in the fermentation process. In few biogas plants, municipal bio-waste from separate collection and green waste from gardens and parks were treated, but only the bio-waste is used in the step of digestion. After the fermentation process, the digestate is often combined with the green cuts within the post-composting process. Thus, the measured emissions of post-composting processes based on the treated waste at the facility in total.

In addition to the measured GHG emissions of the AD plants, further assumptions to calculate the GHG performance were considered.

Emissions from CHP

Due to the fact that not all CHP units were measured, an average emission value for the CHP is assumed. According to measurements of gewitra (personal communications), the median of CH_4 and N_2O emissions of 161 measured CHP units in the range from 300 to 1,000 kW_{el} were determined with 1,760 g CH_4 per ton of bio-waste and 2.1 g of N_2O per ton of bio-waste treated at the facility. Considering the emission factors [1] for N_2O (298) and CH_4 (25), a GWP of 44.6 kg CO_2-eq per ton of bio-waste was estimated for all CHP units.

The energy demand of the investigated biogas plants was determined according to the data of plant operators. It was estimated to cover the electricity demand by using external electricity from the grid. The electricity production in Germany in 2011 produced in average 559 g CO_2-eq per kWh_{el} [20].

Electricity Production

The electricity production from biogas replaces fossil fuels and can be considered as credit [21]. The amount of credit for the electricity production depends on the amount of produced electricity referring to the data of plant operators. The electricity mix of Germany in 2011 with 559 g CO_2-eq per kWh_{el} [20] was assumed to calculate the credit of electricity production.

Heat Utilization

The exhaust heat of electricity generation in CHP units can - if used - substitute heat production based on fossil fuels [21]. The avoided GHG emissions of fossil heat supply by providing heat for external utilization (e.g., district heating, drying process) was stated as heat credits. The amount of heat credit may vary depending on the amount of heat and type of fossil heat, which is replaced in the specific case. With regard to the substitution of fossil heat, an average of the specified external heat mix of 291 g CO_2-eq per kWh_{th} [21] was used to calculate the heat credits.

Digestate - fertilizer and Humus Effects

Depending on the kind of digestate, respectively, the kind of treatment of the fermentation residues (e.g., with/without separation, with/without post-composting after fermentation process), different utilization pathways of digestate have been considered. According to the kind of digestate (finished compost, fresh compost, liquid fermentation residues, solid digestate), different GHG emissions can be saved and considered to the GHG balances as credits (Table 1). Referring to the kind of digestate, the following credits were determined: substitution of mineral fertilizer (nitrogen, phosphorus, potassium), substitution of peat (only in case of finished compost), humus accumulation (carbon-sink), and humus reproduction (i.e., for maintaining soil fertility).

According to the nutrient content (i.e., nitrogen, phosphorus, potassium amounts) of investigated digestates, the production of mineral fertilizer can be substituted and is stated in GHG balances as credit. The following emission factors for the production of mineral fertilizer were assumed according to [22]: 6.41 kg CO_2-eq per kg nitrogen (N), 1.18 kg CO^2-eq per kg phosphorus (P_2O_5), and 0.663 kg CO_2-eq per kg potassium (K_2O).

Humus effects of digestate at investigated AD plants were considered if applied on agricultural land. To evaluate the humus effects of fermentation residues, estimations according to [23] were used. That means, for the amount of finished compost, 20% substitution of peat and 80% agricultural use, thereof 20% of humus accumulation and 80% of humus reproduction was assumed. For the scenario of humus, reproduction was stated - in contrast to [23] - that the substitution of straw is considered and the credits for the fermentation of straw with recirculation of the digestate can be estimated. The humus reproduction (i.e., for maintaining soil fertility) of digestates depends on the content of dry matter and organic dry matter as well as the degrading stability of organic dry matter. Data regarding the humus reproduction of digestate from AD based on bio-waste are not available. The humus reproduction of digestate of investigated AD plants was calculated. The characteristics (e.g., dry matter, organic dry matter, amount of nutrients especially nitrogen) of each digestate were determined based on the

1-year certificate of digestate referring to the quality assurance of the Federal Compost Association.

According to the kind of digestate, the substitution effect compared to straw was analyzed. Therefore, the amount of straw was calculated which might be used for biogas production if the application of digestate on agricultural land is assumed. Differed to the kind of digestate, the amount of straw per ton of digestate (fresh matter) was calculated as follows: 2.11 (finished compost), 1.82 (fresh compost), 0.91 (digestate with post-composting), and 0.15 (liquid digestate). The electricity production of the assumed biogas production due to the fermentation of straw was considered as credit for humus reproduction of digestate.

The substitution of peat was estimated only in case of finished compost. According to the assumptions in [24], 1 kg dry peat (respectively, 2 kg fossil carbon dioxide) is replaced by 1 kg compost (organic dry matter). Referring to the humus accumulation (carbon sink) of composted digestate, the amount of organic carbon (Corg) as published in [23] was assumed as follows: 21.6 kg Corg per ton of digestate for fresh compost and 64.5 kg Corg per ton of digestate for finished compost. In consideration of the stoichiometric ratio of Corg relative to CO_2, 1 kg Corg can fix 3.7 kg CO_2.

Application of Digestate

The application of digestate on agricultural land can cause N_2O emissions as well as NH_3 emissions [25]. With respect to the NH_3 emissions, it was assumed that 1% of the NH_3 is converted to N_2O emissions [1].

RESULTS AND DISCUSSION

GHG Emissions

Various fermentation processes such as wet fermentation, dry fermentation, and batch fermentation were analyzed according to the emission situation. The results show that the emissions are dominated not by the kind of the fermentation process or the technology but by the manner of plant operation.

Figure 1 shows the measured emissions of CH_4, N_2O, and NH_3 (converted to carbon dioxide equivalents) of the investigated AD plants. The range of determined plant emissions varied between 40 and 320 kg CO_2-eq per ton of bio-waste. The detailed presentation on the type of GHGs shows that the CH_4 emissions - except for plant no. 6 - dominate the indicated GHG equivalents of biogas facilities.

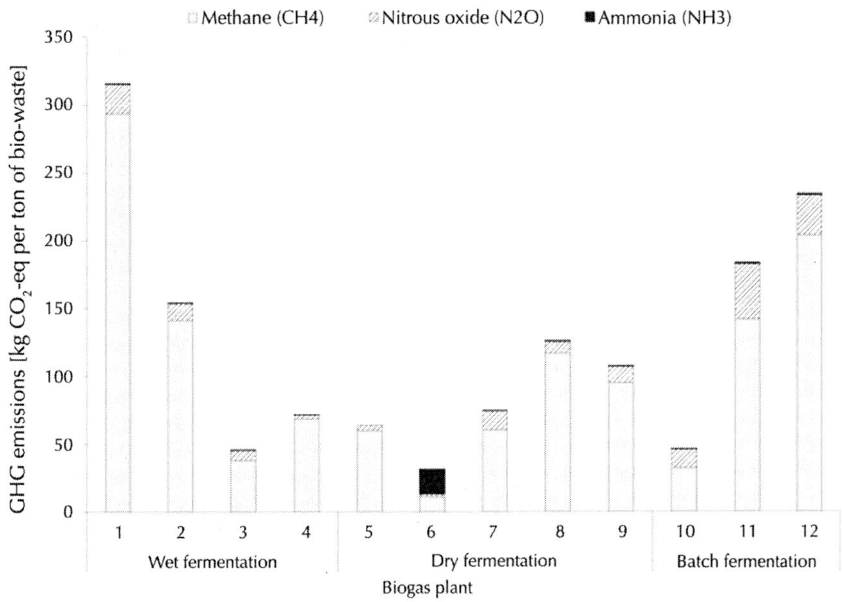

Figure 1: GHG emissions of the investigated biogas facilities (bio-waste) differed to the kind of GHG emission. The measured emissions of investigated AD plants are presented in kg CO_2-eq per ton of bio-waste differed to the kind of GHG emission and kind of fermentation process. Methane and nitrous oxide emissions = direct GHG emissions, ammonia = indirect GHG emissions.

Important sources of GHG emissions were identified. The component-specific GHG emissions of the bio-waste digestion plants are presented in Figure 2.

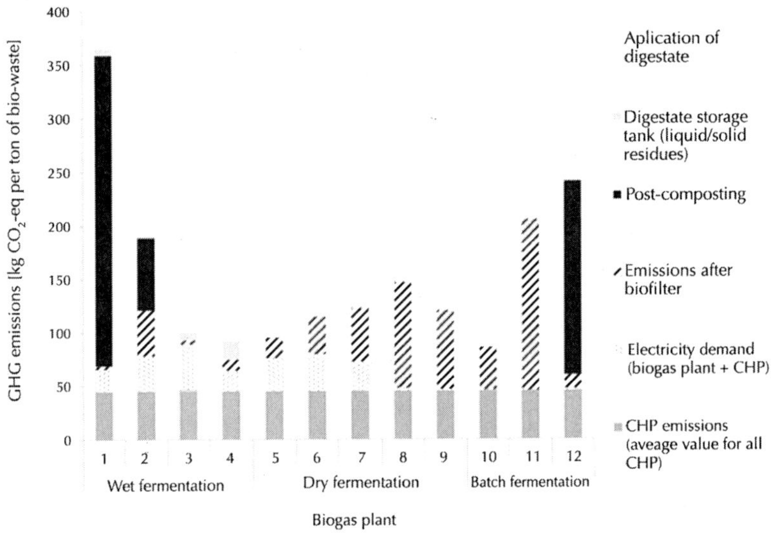

Figure 2: GHG emissions of bio-waste digestion plants differed to kind of plant components. The GHG emissions are presented in kg CO_2-eq per ton of bio-waste differed to the kind of fermentation process (wet, dry, or batch fermentation) and the main emission sources within the process chain.

Especially, the inadequate aeration directly after fermentation (in order to interrupt the methanogenic activity) processes as well as unaerated or less aerated post-composting processes caused extremely high GHG emissions (see plant no. 1, no. 2, or no. 12). In case of some of the investigated biogas plants, the emissions of post-composting are summarized in the amount of 'emissions after bio-filter' (e.g., AD plant no. 10). The overall emissions of AD plant no. 10 was quite low because all parts of the fermentation and post-composting process were totally encapsulated.

Furthermore, AD plant no. 6 showed higher NH_3 emissions due to the drying of digestate at higher temperature and higher pH value. In this case, the existing downstream acidic scrubber was out of operation during the measurements. The operation of the bio-filters can also be problematic; extremely wet bio-filters for example can cause additional CH_4 production as observed at AD plant no. 8.

Finally, on almost all AD plants, emission sources were identified whose intensity can be reduced if the state-of-the-art treatment technology was used (e.g., acid scrubber before bio-filter, aeration of post-composting). The results show that the open storage of fermentation residues (with or without separation step) should be avoided. In addition to unaerated post-composting processes and open storage of active material (e.g., solid digestate), the CHP was one of the most important sources of CH_4.

According to the measured residual gas potential of digestate, a wide range from 4 to 23% was determined. Ten of 12 samples of digestate of investigated AD plants showed a relative residual gas potential higher than 10%. A high relative residual gas potential means insufficient fermentation of the substrate. The residual gas potential of bio-waste digestion achieved the same range as agricultural AD plants which were operated as single-stage processes, whereas in comparison to agricultural biogas plants with multi-stage process, the determined CH_4 potential of fermentation residues from bio-waste digestion provides basically higher values. Table 4 shows the gas potential of the investigated bio-waste plants compared to the gas potential of agricultural biogas plants as published in [17]. According to [17] where agricultural AD plants were investigated, discontinuous systems (batch) and single-stage systems have shown the highest residual gas potential. Moreover, multi-stage systems of agricultural AD plants achieved less than half of the residual gas potential of single-stage plants [17]. The results of [17] stated that single-stage processes achieve higher residual gas potential due to their generally shorter retention time. With respect to the investigated bio-waste AD plants hydraulic retention times (HRTs) ranged from 1 to 4 weeks. However, due to a great variability of other process parameters, the results do not give a clear answer regarding the estimation that lower HRT corresponds to lower gas potential (see Tables 5 and 6).

Table 4: Investigated AD plants differed to kind of digestate and considered GHG credits (marked with 'x')

	Kind of digestate	Solid digestate				Liquid digestate	
		Substitution of mineral fertilizer	Substitution of peat	**Humus accumulation**	**Humus reproduction**	**Substitution of mineral fertilizer**	**Humus reproduction**
1	Fresh compost	x	-	x	x	-	-
2	Finished compost	x	x	x	x	-	-
3	Solid digestate (separated)	x	-	-	x	-	-
4	Liquid digestate	x	-	-	x	x	x
5	Separated digestate without post-composting	x	-	-	x	x	x
6[a]	Solid digestate	x		-	x	-	-
7	Fresh and finished compost	x	x	x	x	-	-

8	Fresh and finished compost	x	x	x	x	x	x
9[b]	Finished compost	x	x	x	x	x	x
10	Finished compost	x	x	x	x	-	-
11	Fresh and finished compost	x	x	x	x	-	-
12	Finished compost	x	x	x	x	-	-

[a]Assumption according to plant no. 5 (no data available). [b]Assumption according to plant no. 8 (no data available).

Daniel-Gromke et al.

Daniel-Gromke et al. Energy, Sustainability and Society 2015 5:3, doi: 10.1186/s13705-014-0032-6

Table 5: Residual gas potential in percent related to the methane production

	Bio-waste AD plants 38°C	Agricultural AD plants 37°C	
		Single-stage	Multi-stage
Average	14.1	9.7	5.1
Min	3.6	3.2	1.2
Max	23.4	21.8	15

Investigated bio-waste AD plants in comparison to agricultural AD plants according to Weiland et al. [17] and temperature level for determination of gas potential.

Daniel-Gromke et al.

Daniel-Gromke et al. Energy, Sustainability and Society 2015 5:3, doi: 10.1186/s13705-014-0032-6

Table 6: Hydraulic retention time and residual gas potential of investigated AD plants

Plant no.	HRT in days[a]	Relative residual gas potential in percent[b]		Kind of fermentation[c]	Mode of operation	Separation of digestate
		Minimum	Maximum			
1	8	11	15	Wet	Multi-stage	Yes
2	20	19	19	Wet	Multi-stage	Yes
3	17	-	-	Wet	Single-stage	No
4	25	-	-	Wet	Multi-stage	No
5	25	-	-	Dry	Single-stage	Yes
6	21	15	21	Dry	Single-stage	No
7	28	4	11	Dry	Single-stage	No
8	14	23	23	Dry	Single-stage	No
9	21	17	17	Dry	Single-stage	No
10	28	12	17	Batch	Single-stage	No
11	21	-	-	Batch	Single-stage	No
12	21	6	6	Batch	Single-stage	No

[a]Hydraulic retention time in days. [b]Residual gas potential of digestate according to the input material (fresh matter based). [c]Wet = wet fermentation, dry = dry fermentation, batch = batch system (discontinuous).

Daniel-Gromke et al.

Daniel-Gromke et al. Energy, Sustainability and Society 2015 5:3, doi: 10.1186/s13705-014-0032-6

GHG Balances

The overall GHG balance of the investigated AD plants depends on the measured GHG emissions on the one hand (see 'GHG emissions') and on the credits for the generated products (e.g., combined heat and electricity from biogas; fertilizer and humus supply from fermentation residues) on the other hand. The calculated GHG credits according to the AD plant concept are presented in Figure 3.

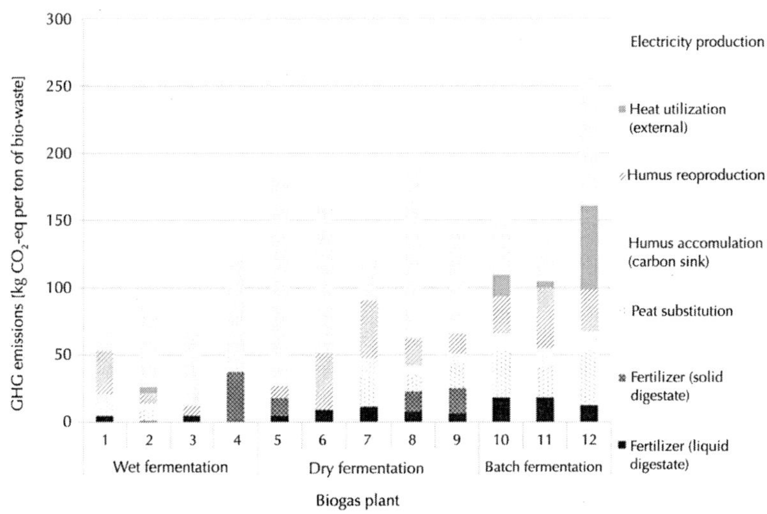

Figure 3: GHG credits of investigated bio-waste digestion plants.GHG credits depend on the amount of energy production or heat utilization (substitution of fossil electricity/fossil heat production) as well as the kind and amount of digestate (substitution of fertilizer, substitution of peat, humus effects).

Finally, the highest amount of GHG credits of humus reproduction can be expected from composted digestate. In general, the following order of humus reproduction can be assumed: post-composted digestate (finished and fresh compost) > solid digestate > liquid digestate. In case of finished compost, additional GHG credits for the substitution of peat (by application in soil producing facilities, e.g.) can be considered.

If external heat (generated by the electricity production of CHP unit) is utilized, credits for avoided fossil heat production optimize the GHG balances as well (see plant no. 12). Nevertheless, in most cases (besides plants nos. 1, 7, 10, and 11), the credit for the electricity production based on biogas which was given for the substitution of fossil fuels dominates the GHG credits.

The total range of GHG balances (including credits) varied between −49 and 323 kg CO_2-eq per ton of bio-waste due to different plant concepts and measured emissions (see Figure 4).

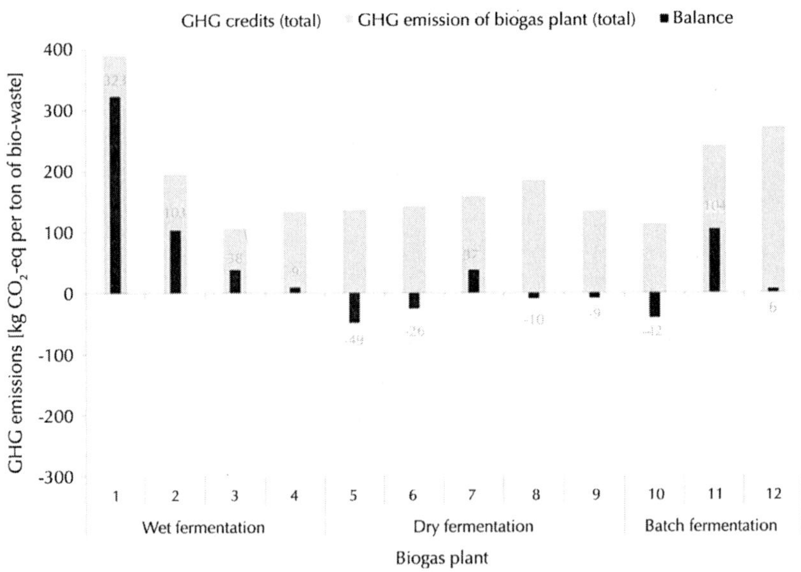

Figure 4: Total GHG balance of bio-waste digestion plants with GHG emissions of AD plant and GHG credits. The balance as a result of total GHG emissions of AD plant and total GHG credits (*black column*).

Moreover, the emissions of each component have been set in relation to the amount of produced electricity in order to get an emission value according to the energy output (g CH_4/kWh_{el}). Compared to an assumed electricity mix in Germany (559 g CO_2-eq per kWh_{el} according to [20]), 8 of 12 AD plants show even lower values.

Overall Discussion of Results Gained in this Study

The problem of increased emissions is not the anaerobic process itself, but a non-optimal after-treatment of the digestate. In general, the emission situation is not uniform; the plants show very different emission rates. The total emissions from AD plants no. 3, no. 6, and no. 10 were quite lower than the remaining. However, even those plants showed considerable potential for optimization. The best overall result of the analyzed AD plants belonged to a biogas facility with no external heat utilization and below-average credits for digestate. It can be stated that all investigated biogas facilities showed potential for optimization. Often, there are no incentives for a sufficient utilization of waste with respect to high CH_4 yields or reduction of emissions, due to the fact that the running costs of waste facilities has to be financed by the producers of the waste paying for the waste disposal. Moreover, there are no strict regulations to avoid uncontrolled emissions as for agricultural biogas plants for energy crops and for co-digestion of waste. Therefore, waste treatment plants show relevant potentials for optimization.

AD plant no. 12 showed that very high emissions can be covered by a very good energy concept combined with a good utilization of fermentation residues. The bad overall GHG balance of AD plant no. 1 evidences how certain factors may interact the GHG performance negatively. In this case, extremely high emissions occurring from the post-composting process and very low electricity generation caused high GHG emissions in total. Inadequate digestion of the substrate caused not only low gas production, respectively, electricity generation but also high emissions during the post-composting process of digestate.

Regarding the GHG credits, the highest importance of an efficient fermentation had the production of energy. A high share of electricity generation led to high GHG credits. As far as the utilization of exhaust heat of electricity production was possible, it had also a positive influence on the GHG performance of the AD plant. Moreover, the use of digestate showed positive effects on the GHG balances. In addition to the nutrient effect through the utilization of the fermentation residues as a fertilizer (substitution of mineral fertilizer), GHG emissions can be saved due to the humus effect of digestate. Especially, composted digestate like fresh and finished compost contributed to the humus accumulation (carbon sink) and the humus reproduction of digestate. Compared to the production of fresh or finished compost digestate without post-composting process, which is used within the agriculture directly, less GHG credits were given. However, the risk of high emissions during the post-treatment of the fermentation residues was avoided.

The following measures are able to reduce GHG emission of bio-waste digestion: intensive aeration of the (solid) digestate after fermentation; gas-tight storage tank for fermentation residue and integration into biogas utilization; avoidance of any open storage of digestate and fermentation residues; and small, aerated compost windrows combined with sufficient structural materials and frequent turnover as well as the use of acidic scrubbers in front of the bio-filter.

With respect to the development of methodology of emission measurements and the standardization of procedure for the determination of emissions on biogas plants, further investigations are necessary. Further scientific data about the current emission situation and the ongoing development as well as reliable measurement methods are required to determine the CH_4 emissions from the plants in operation today. In this regard, the reliable measurement of stationary and diffuse emission sources is of high importance. Uncertain are the emissions sources that are not coupled to the gas system of the plant, but still cause GHG emissions as stated in [10]. As one example, no assessment of emissions from pressure relief valves could be carried out as part of this study. Concerning the emissions, the treatment and evaluation of temporary occurring emissions caused by certain operational conditions are still unclear. Moreover, the further development of ecological assessment of biogas pathways with respect to the humus effects of digestate in comparison to other pathways is of great importance.

CONCLUSIONS

Based on the emission measurements, significant sources of emissions were identified. The results show that GHG emissions can be minimized, if the technology and operation of the plant are adjusted accordingly. Basically, the kind of operation of the plant and the handling of digestate determine the amount of GHG emissions. The overall GHG balances of the investigated AD plants depend on the measured emissions as well as the amount of credits for the generated products (e.g., combined heat and electricity from biogas; fertilizer and humus effects from fermentation residues). The consideration of GHG credits can optimize the overall GHG performance of the biogas facilities.

AUTHORS' CONTRIBUTIONS

JDG carried out the total GHG balances including GHG credits and conceived of the papers structure. JL and CK conducted the emission analysis regarding the biogas plant and the components. JL, CK, and JDG described the mitigation potential. VD was responsible for the survey of plant operators and the assessment of these data. All authors drafted, read, and approved the final manuscript.

ACKNOWLEDGEMENTS

The DBFZ investigated the emission situation of biogas plants based on bio-waste in Germany within a research project in cooperation with gewitra mbH and Dr. Reinhold & Kollegen during 2009 to 2012. Gewitra mbH led the measurements on AD plants and post-composting processes. Dr. Reinhold & Kollegen Potsdam was involved into evaluation of humus effects of digestate at investigated AD plants in strong cooperation with our colleague Dr. Walter Stinner. The authors would like to thank Dr. Carsten Cuhls, Birte Mähl, and Dr. Jürgen Reinhold for their cooperation as project partners as well as the German Federal Ministry for the Environment, Nature Conservation, and Nuclear Safety for the financial support of this project.

REFERENCES

1. (2007) Climate change 2007, Fourth Assessment Report (AR4) of WG 1 Intergovernmental Panel on Climate Change (IPCC). Cambridge University Press, Cambridge and New York.
2. Vogt R (2008) Basis data for GHG balances of process chains of biogas and compilation of new balances. German edition: Basisdaten zu THG Bilanzen für Biogas-Prozessketten und Erstellung neuer THG-Bilanzen, Report on behalf of E.ON Ruhrgas AG. Institut für Energie- und Umweltforschung Heidelberg GmbH, Heidelberg.
3. Müller-Langer F, Rönsch S, Weithäuser M, Oehmichen K, Scholwin F, Höra S, Scheftelowitz M, Seiffert M (2009) Economical and ecological assessment of substitutes for natural gas based on

energy crops. German edition: Ökonomische und ökologische Bewertung von Erdgassubstituten aus nachwachsenden Rohstoffen. Deutsches Biomasseforschungszentrum gGmbH, Leipzig.

4. Liebetrau J, Clemens J, Cuhls C, Hafermann C, Friehe J, Weiland P, Daniel-Gromke J (2010) Methane emissions from biogas producing facilities within the agricultural sector. Eng Life Sci 10:1-5
5. Liebetrau J, Reuschel C, Clement J, Friehe J, Weiland P (2013) Analysis of greenhouse gas emissions from 10 biogas plants within the agricultural sector. Water Sci Technol 67(6):1370-1379 doi:10.2166/wst.2013.005
6. Petersson A (2012) The Swedish Voluntary system for control of methane emissions. http://www.iea-biogas.net/case-studies.html?file=files/daten-redaktion/download/case-studies/Case%20Study%20Sweden.pdf. Accessed 23 Jul 2013
7. Flesch T, Desjardins R, Worth D (2011) Fugitive methane emissions from an agricultural biodigester. Biomass Bioenergy 35(9):3927-3935
8. Woess-Gallasch S, Enzinger P, Jungmeier G, Padzinger R (2007) Greenhouse gas emissions from biogas plants. Report for Landesenergieverein Steiermark, Graz.
9. Amon B, Kryvoruchko V, Amon T, Zechenmeister-Boltenstern S (2006) Methane, nitrous oxide and ammonia emissions during storage and after application of dairy cattle slurry and influence of slurry treatment. Agric Ecosyst Environ 112:153-162
10. Dumont M, Luning L, Yildiz I, Koop K (2013) Methane emissions in biogas production. In: Wellinger A, Murphy J, Baxter D (eds) The biogas handbook - science, production and application, Series in Energy, Cambridge. pp 248-266
11. DESTATIS (2013) Abfall-Daten. Statistisches Bundesamt. https://www.destatis.de/DE/ZahlenFakten/GesamtwirtschaftUmwelt/Umwelt/UmweltstatistischeErhebungen/Abfallwirtschaft/Tabellen/Abfallbilanz2011.pdf?__blob=publicationFile. Accessed 20 Mrch 2013
12. Fricke K, Heußner C, Hüttner A, Turk T, Bauer W, Bidlingmaier W (2013) Vergärung von Bio- und Grünabfällen, Teil 1:

Ausbaupotenzial bei der Vergärung von Bio- und Grünabfällen. Müll Und Abfall 12:628-635

13. Daniel-Gromke J, Rensberg N, Schüch A (2014) Ein Überblick - Potenziale, Techniken und Perspektiven von zukünftigen Substraten im Biogassektor mit Fokus auf die Bioabfallvergärung. In: Nelles M (ed) DBFZ-Jahrestagung 2014 Bioenergie. Vielseitig, sicher, wirtschaftlich, sauber?! 1.-2. Oktober 2014, Deutsches Biomasseforschungszentrum gGmbH, Leipzig. ISSN 2199-9384
14. Hashmonay RA, Varma RM, Modrak MT, Kagann RH, Segall RR, Sullivan PD (2008) Radial plumemapping: a US EPA test method for area and fugitive source emission monitoring using optical remote sensing. In: Kim YJ, Platt U (eds) Advanced environmental monitoring, Springer, Cham, Switzerland. pp 21-36
15. Liebetrau J, Daniel-Gromke J, Krebs C (2012) Emissionen aus Vergärungsanlagen. Erfahrungen aus dem landwirtschaftlichen Bereich und aus der Abfallbehandlung. In: Immissionsschutz. Aktuelle Entwicklungen im anlagenbezogenen Planungsprozess und Immissionsschutz. Band 3, Neuruppin, S. 337-351 – ISBN 978-3- 935317-90-0
16. Oechsner H (2013) Determination of residual gas- /residual methane potential. In: Liebetrau J, Pfeiffer D, Thrän D (eds) Method handbook - biogas as part of the funding programme "Biomass for Energy", vol 7, Report commissioned by the Federal Ministry of Environment, Nature Conservation and Nuclear Safety, Deutsches Biomasseforschungszentrum gGmbH, Leipzig.
17. Weiland P, Gemmeke B, Rieger C, Plogsties V, Vogtherr J, Kissel R, Bachmaier H, Schumacher B (2009) Biogas-Messprogramm II – 61 Biogasanlagen im Vergleich. Fachagentur Nachwachsende Rohstoffe e. V. (ed), Gülzow
18. Reinhold G, Gödeke K (2011) Restgaspotential in landwirtschaftlichen Biogasanlagen. Biogas – Fachtagung Thüringen. http://www.tll.de/ainfo/pdf/bior0111.pdf. Accessed 15 Jan 2015
19. (2006) VDI 4630, Vergärung organischer Stoffe, Substratcharakterisierung, Probenahme, Stoffdatenerhebung. Gärversuche, Norm, Düsseldorf.
20. (2012) Entwicklung der spezifischen Kohlendioxid-Emissionen des deutschen Strommix 1990-2010 und erste Schätzungen

2011. Federal Environment Agency (Umweltbundesamt), Dessau-Roßlau.

21. Thrän D, Pfeiffer D (2011) Method handbook - Material flow-oriented balancing of greenhouse gas effects - methods for the determining of technology indicators, production costs, and greenhouse gas effects of projects as part of the funding programme "Biomass for Energy", vol 1. Nature Conservation and Nuclear Safety. Deutsches Biomasseforschungszentrum gGmbH, Leipzig.
22. (2010) Leitfaden Nachhaltige Biomasseherstellung. Bundesanstalt für Landwirtschaft und Ernährung (BLE), Bonn.
23. Knappe F, Vogt R, Lazar S, Höke S (2012) Optimierung der Verwertung organischer Abfälle, Report commissioned by the Federal Environment Agency (Umweltbundesamt). Institut für Energie- und Umweltforschung Heidelberg GmbH, Heidelberg.
24. Vogt R, Knappe F, Giegrich J, Detzel A (2002) Ökobilanz Bioabfallverwertung. IFEU Institut für Energie- und Umweltforschung Heidelberg GmbH, Report commissioned by the German Federal Environmental Foundation. Deutsche Bundesstiftung Umwelt DBU, Berlin.
25. Wulf S, Maeting M, Clemens J (2002) Application technique and slurry co-fermentation - effects on ammonia, nitrous oxide, and methane emissions after spreading. J Environ Qual 31:1795-1801 doi:10.2134/jeq2002.1795.

Chapter 8

Monitoring Land-Use Change in Nakuru (Kenya) Using Multi-Sensor Satellite Data

Kenneth Mubea[1] and Gunter Menz[1, 2]

[1]Remote Sensing Research Group (RSRG), University of Bonn, Bonn, Germany

[2]Center for Remote Sensing of Land surfaces (ZFL), University of Bonn, Bonn, Germany

ABSTRACT

Recently land-use change has been the main concern for worldwide environment change and is being used by city and regional planners to design sustainable cities. Nakuru in the central Rift Valley of Kenya has undergone rapid urban growth in last decade. This paper focused on urban growth using multi-sensor satellite imageries and explored the potential benefits of combining data from optical sensors (Landsat, Worldview-2) with Radar sensor data from Advanced Land Observing Satellite (ALOS) Phased Array type L-band Synthetic Aperture Radar (PALSAR) data for urban land-use mapping. Landsat has sufficient spectral bands allowing for better delineation of urban green and impervious surface, Worldview-2 has a higher spatial resolution and facilitates urban growth mapping while PALSAR has higher temporal resolution compared to other operational sensors and has the capability of penetrating clouds irrespective of weather conditions and time of day, a condition prevalent in Nakuru, because it lies in a tropical area. Several classical and modern classifiers namely maximum likelihood (ML) and support vector machine (SVM) were applied for image classification and their performance assessed. The land-use data of the years 1986, 2000 and 2010 were compiled and analyzed using post classification comparison (PCC). The value of combining multi-temporal Landsat imagery and PALSAR was explored and achieved in this research. Our research illustrated that SVM algorithm yielded better results compared to ML. The integration of Landsat and ALOS PALSAR gave good results compared to when ALOS PALSAR was classified alone. 19.70 km^2 of land changed to urban land-use from non-urban land-use between the years 2000 to 2010 indicating rapid urban growth has taken place. Land-use information is useful for the comprehensive land-use planning and an integrated management of resources to ensure sustainability of land and to achieve social Equity, economic efficiency and environmental sustainability.

INTRODUCTION

Quantitative urban studies are becoming increasingly important for planners knowing that in the year 2015 more than half the global population will be residing in cities [1]. Suitable urban planning ought to be a top priority for future development but unfortunately sound planning has not taken place especially in many African cities as heavy rural-urban migration continues to cause cities to expand at uncontrollable rates [2]. As a consequence, the urban population in Africa is increasing at a much faster rate than in the rest of the world, contributing to the augmentation of the existing problems such as unsuitable land-use [3]. The concentration of population in cities comprises as much as 60% of the total population in most countries. In these immense urban settlements the environmental and social consequences are sometimes disastrous [4].

Large cities in Africa such as Nakuru have experienced a fast growth rate of 13.3% between 1990 and 2006 [5]. The magnification has been attributed to a number of factors, mainly the aperture of the new Naivasha-Nakuru road, which links the megacity of Nairobi. Post-election violence is verbally expressed to be one of the contributing factors, since many displaced people from neighboring towns migrated to Nakuru as a safe shelter.

The main consequences in these African cities include; urban sprawl, unsuitable land-use, inadequate transportation systems, air and water pollution, depletion of natural resources, collapse of public services, proliferation of epidemics, and other negative environmental and social effects. The changing of surrounding area due to uncontrolled city development and city residents ever increaseing demand for energy, food, goods and other options is behind the deterioration of local and localized environment which is harmful the basic environment solutions and biodiversity. Solutions of problems linked to unsustainable urban development in African cities are manifold and complex and requires an integrated spatially explicit approach. Such an integrated urban planning approach needs to recognize and anticipate urban dynamics and their consequences [6].

Remote sensing techniques have been valuable in mapping urban land-use pattern as well as data sources which aid in the analysis and modeling of urban growth and land-use change [7]. Remote Sensing offers spatially coherent data sets that cover large areas with both high spatial detail and high temporal frequency. These data characteristics are necessary for land-use monitoring, which is an essential element of socio-ecological studies. As urbanization occurs, changes in land-use accelerate and land making up the natural resource base such as forests and agricultural land, leading to modification and conversion of existing land-uses [8]. This is referred to as land take [9].

In order to assess the land-use in Nakuru town, it is best to evaluate spatial dimension of the imageries. The spatial dimension of remote sensing images as assessed by image texture contains information on local spatial structure and variability of land-cover categories, and can raise land-use classification accuracies in heterogeneous urban landscapes [10]. Texture information has improved classification accuracy for optical sensors such as the Satellite Pour l'Observation de la Terre (SPOT) High Resolution Visible (HRV) sensor, Landsat TM, Multispectral Electro optical Imaging Scanner (MEIS-II), and airborne multispectral sensors. The optimal window size for texture measurements is highly dependent on the image pixel size and the land cover characteristics [11]. Normally, window size should be large enough to include the entire texture pattern, and at the same time small enough to include only one land-cover type [12,13].

Very high resolution SAR sensors are playing an increasingly important role in urban remote sensing due to their ability to operate day and night through cloud cover, recent improvement in data availability and spatial resolution [14]. Many studies have focused on the frequency and polarimetry of SAR data in land cover classification [15], whereas SAR image texture is found helpful in improving map accuracy, particularly for urban and forest categories [16].

Multi-temporal SAR images have proven to be useful in urban, forest, and agriculture land-cover classification [17]. Recent studies report that the fusion of optical and SAR data is useful due to their capability to distinct different features. Optical images contain information on surface reflectance and emissivity characteristics, while SAR images capture the structure and dielectric properties of the Earth surface materials [18]. Land-cover types that are impossible to separate in optical images might be distinguishable with SAR images and vice versa because of the complementary information contained in the two spectral domains [19].

Many approaches employing both optical and SAR images have been explored for land-cover classification [20,21]. The results from integrating optical and SAR sensors are often significantly higher than those obtained from using an individual sensor, particularly for certain land-use types, such as urban, agriculture and wetlands.

The study of land-use and land-cover changes is essential not only for land-use management but also in detecting environmental change and in formulating sustainable development strategies [22]. Accurate information on land-use changes is needed for documenting urban growth, making policy decisions and improving land-use planning. Information concerning land-use changes is also required for predictive modeling [23].

The aim of this research was to perform different land-use classification using multi-sensor data and to monitor land-use change. To achieve this optical and SAR datasets were obtained. Optical datasets included: Landsat of 1986, 2000 and 2010; Worldview-2 of 2010; and ALOS PALSAR imagery of the year 2010. Image classification was performed on resampled datasets at 12.5 meters and 28.5 meters. Several image classification algorithms were explored and their performance evaluated using accuracies measured such as overall accuracies and kappa coefficient values. Support vector machine (SVM) performed better compared to maximum likelihood classifier (ML). Thus combining images from optical and SAR yielded better results in image classification. Landuse change monitoring was achieved by comparing the change between two corresponding years and the result given as a percentage.

THE STUDY AREA

Nakuru municipality lies in Central Rift Valley in Kenya between latitudes 0°15›S and 0°31›S, and longitude 36°00›E and 36°12›E, with an average altitude of 1859 meters above sea level, covering an area of 290 km^2 (Figure 1). Within Nakuru municipality is Nakuru town, and Lake Nakuru National Park. The Lake Nakuru National Park is a tourist attraction of great economic value for the country with Lake Nakuru being one of the largest bird sanctuaries in the world with the flamingo and pelican bird species. The town of Nakuru is located 160 km North West of Nairobi along the twin east-west railroad transport route from Mombasa to Kampala and is the fourth largest city in Kenya following Nairobi, Mombasa and Kisumu. The administratively defined city has land-uses divided into urban use, agriculture, range land and remnants of evergreen tropical forests. Such land-uses are similar to other cities in Kenya and sub Saharan Africa [2].

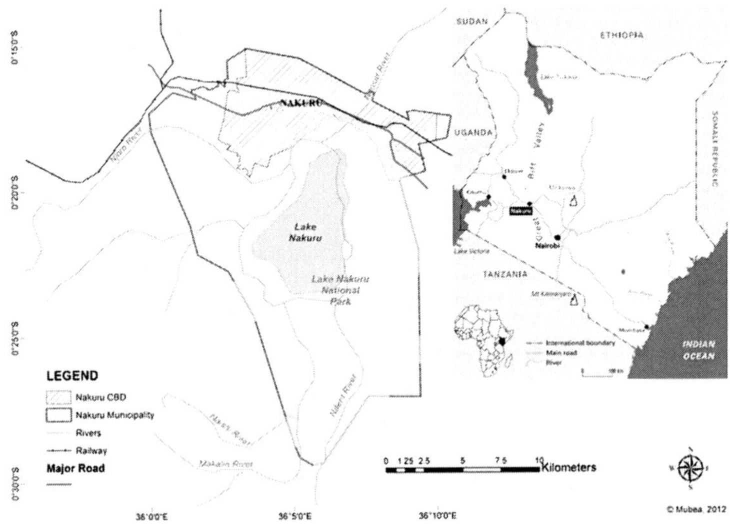

Figure 1: Location of Nakuru in the central rift valley of kenya (source: World Resource Institute).

Urban Growth Rates

The urban pattern of Nakuru town and its environs are characterized by intense urban pressures, first along the main highways and through the development of sub urban areas. Nakuru population has been growing at the rate of 5.6% per annum. From a population of 38,181 in 1962, the population reached 47,151 in 1969 [24], 92,851 in 1979 [25], 163,927 in 1989 [26], and 289,385 in 1999 [27]. By the year 2015, the population is projected to rise to 760,000, which is approximately 50% above the present levels [28].

As a consequence of this population development, urban sprawl has led a negative impact on infrastructure and the sustainability of cities [29].This has led to; spontaneous slum developments, large increase in the number of street children, unemployment and high rates of crime, strain on existing urban infrastructure and services, and many urban areas are faced with severe environmental problems for example air pollution.

With respect to institutional arrangements, the local governments, entrusted with the provision of urban basic infrastructure, have been unable to perform as a result of administrative problems and lack of capability [29].

In order to address urban growth in Nakuru, this research utilized a multi-sensor approach. Landsat, Worldview-2 and ALOS-PALSAR data were used to better delineate urban land-use and its change within Nakuru municipality.

UTILITY OF MULTI-SENSOR SATELLITE DATA

Urban landscapes remain one of the most challenging environments to be analyzed from remotely sensed data. They are complex, featuring spatial and spectral heterogeneity and are composed of multi-fold artificial and natural surface types [30]. In a comparative analysis of urban reflectance, [31] concluded that sensors with similar spectral and radiometric properties to Landsat Thematic Mapper (TM) principally comprise a threedimensional spectral feature space. In such a feature space, linear combinations of spectral end members (high albedo, vegetation and low albedo) cover approximately 98% of the variance contained in an urban scene [32].

Challenges of urban mapping can be solved using more temporal information and data from different sensors. Integrating multi-temporal information helps distinguish urban from non-urban surfaces as urban spectral responses are largely persistent over time compared to non-urban surface phenology [32].

Combining multi-spectral imagery with data from the microwave spectral domain has also proven to be a powerful strategy, especially for monitoring urban areas and for overcoming spectral ambiguities [33,34]. The combined use of synthetic aperture radar (SAR) and multispectral optical imagery has been successfully used to delineate built-up areas [35].

Urban mapping requires the selection of an appropriate scale of observation. This involves a certain trade-off between the richness of detail of very high resolution (VHR) remote sensing imagery and the generalizing nature of moderate to high resolution sensors such as the 30 m of Landsat data [32]. The long-term and future data record is an advantage of Landsat data [36,37] compared to very high resolution sensors. Optical and SAR systems operate in different wavelengths, ranging from visible to microwave.

Landsat sensors include MSS (Multi-Spectral Scanner), TM (Thematic Mapper) and ETM+ (Enhanced Thematic Mapper Plus). Figure 2 illustrates a false color composite Landsat RGB (red, green and blue bands) and panchromatic band of Nakuru municipality. World view-2 was launched on 8thOctober 2009. It is the first comercial eight multispectral bands high resolution satellite (sensor).Figure 3 illustrates Worldview-2 True color bands of the central business district (CBD) of Nakuru.

Radar wavelength has an ultimate influence on the interaction between the electromagnetic wave and the natural medium [38]. As wavelength increases, surface roughness criteria will also change. In general, more surface features will appear smoother at longer wavelengths than at shorter wavelengths.

The capability to penetrate through precipitation clouds or into a surface layer is increased with longer wavelengths. Typically the penetration is half the wavelength and as wavelength increases and frequency decreases, respectively, the penetration becomes higher [39].

Radar is able to receive and process all four combinations of polarizations: HH, HV, VH, and VV. HH stands for horizontal transmit and horizontal receive, VV stands for vertical transmit and vertical receive, HV stands for horizontal transmit and vertical receive, and VH stands for vertical transmit and horizontal receive. HH and VV are referred to as like-polarized because the transmit and receive polarizations are the same, while HV and VH are referred to as cross-polarized because the transmit and receive polarizations are orthogonal to one another. The channels have varying sensitivities to differentiate surface characteristics and properties. For instance, the dynamic range of the like-polarized module is larger than that of the cross-polarized module for urban areas; this is in contrast to the measurement for forested areas, where the dynamic range of the cross-polarized component is larger than that of the like-polarized component [40].

Multiple polarizations help to distinguish the physical structure of the scattering surfaces e.g. HH vs. VV is used in the alignment with respect to the radar, HV is used in the randomness of scattering for vegetation cover, HH and VV phase angle is used for the corner structures and urban land-use mapping.

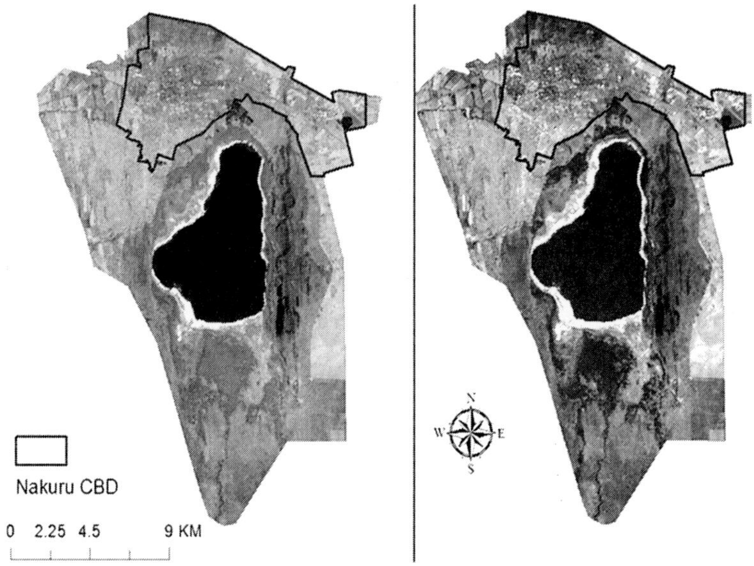

Figure 2: (a) False color composite Landsat tm (path 169, row 60) RGB of 1986 for Nakuru municipality (b) panchromatic band of 1986 for Nakuru municipality.

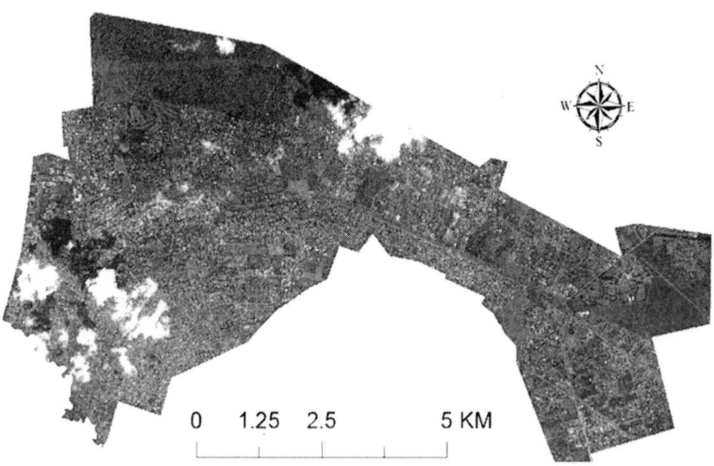

Figure 3: Worldview-2 true color image of year 2010 for Nakuru CBD (extend of CBD see Figure 2).

Past studies showed that radar imagery collected using different polarization may provide different and complementary information about the targets on the surface and thus improve the interpretation of different urban features [40]. The use of polarimetric datasets for urban analysis has been accelerated in recent years with more and more fully-polarimetric SAR systems being in operation for example Terra SAR-X. Several studies demonstrated that polarimetric datasets have great potential for urban applications [38,41].

ANALYSIS

Data

The approach adopted for the analysis of land-use involved: Landsat images for 1986, 2000 and 2010, Worldview-2; and Advanced Land Observing Satellite (ALOS) Phased Array type L-band Synthetic Aperture Radar (PALSAR) image for 2010 were selected for the study area. The ALOS PALSAR was at level 1.5 and polarimetric mode, multi-polarized images, HH Horizontal-Horizontal + HV Horizontal-Vertical + VerticalHorizontal + VV Vertical-Vertical. Nakuru municipality is entirely contained within Landsat TM path 169, rows 60. The Landsat data sets include TM, and ETM + images. Digital elevation model (DEM) at spatial resolution of 90 meters was obtained from the United States Geological Survey (USGS) and used to pre-process the SAR data. Reference data included a topographic map (scale 1:50,000) and ground truth information about land-use from 50 GPS points which were used for classifier training and accuracy assessment. Stratified random sampling was adopted for selecting samples.

Land-Use Classification

Two approaches were implemented for the land-use classification of Nakuru; Nakuru municipality with six landuse classes (namely urban, water, forest, agriculture, barren land and rangeland) and Nakuru Central Business District (CBD) with three land-use classes (namely urban, forest and agriculture). The three classes were selected for Nakuru CBD since it is the center of major activities and infrastructure.

Several factors were considered during the design of categorization scheme such as the major land-use categories within the research area, disparities in spatial rules of the sensors, and the need to always discriminate land-use classes irrespective of seasonal disparities [42].

Image pre-processing steps for the optical datasets were radiometric and geometric correction as illustrated in Figure 4. GPS points were used for image to map registration. UTM 37 South was selected as reference system for the research area. The processing of the SAR data was done using European Space Agency (ESA)

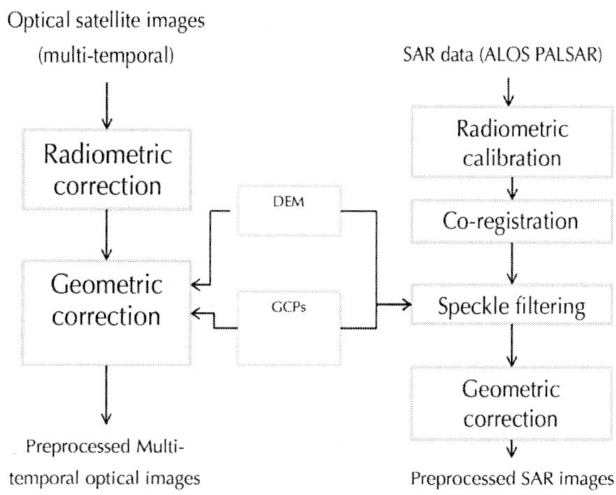

Figure 4: Image preprocessing workflow of the multi-sensor satellite data sets.

NEST SAR Remote Sensing software. Multi-look correction was applied. Radar speckle which appear as grainy "salt and pepper" texture in imagery, was reduced prior to interpretation and analysis using Gamma Map filter with 5 × 5 kernel. The DEM used was an SRTM (Shuttle Radar Topography Mission) and was used to remove relief displacement in SAR data In order to find the optimal spatial resolution scenarios of upscaling and downscaling of the data sets were explored. Upscaling refers to the decrease in spatial resolution while downscaling is the increase in spatial resolution.

Scale has several definitions according to [43] namely: cartographic or map scale; observational or geographic scale that refers to the size or spatial extent of the study; measurement scale or resolution that refers to the minimum size that can be distinguished; and finally operational scale that refers to the scale at which certain processes operate in the environment.

In a nutshell, scale is the spatial and temporal dimension of the research while at the same time; scale also refers to the spatial extent and frequency that a certain phenomenon or process occurs [44]. Additionally, scale shows the level on which the object is understood. Hence scale can either be temporal or spatial scale, and of which both were investigated in this research. Combinations of the reflective spectral bands from images (i.e., stacked vector) were used for classification of the 1986, 2000 and 2010 images. The ratio of HH/HV was used for the case of the ALOS PALSAR imagery which yields best results for urban areas [18,45].

Image processing techniques were applied to the data sets both at spatial resolution of 12.5 meters and 28.5 meters using nearest neighbor interpolation. Therefore the Worldview-2 data was resampled from 0.5 meters (original) to 12.5 meters and 28.5 meters. Landsat was resampled from 28.5 meters (original) to 12.5 meters while ALOS PALSAR was resampled from 12.5 meters (original) to 28.5 meters. The objective was to assess how classification accuracy changes with spatial resolution. Image processing workflow is illustrated in Figure 5.

Training sites representing the land-use classes of interest were collected using the region growing tool in ENVI 4.8. Such sites were homogeneous and extensive to provide excellent statistics. ML was applied to the multi-sensor datasets, and performance assessed using the corresponding confusion matrices.

SVM classification in ENVI 4.8 was applied to all the data sets and its performance assessed using confusion matrices. The SVM classifier has four kernels namely linear, polynomial, Radial Basis Function (RBF) and sigmoid. Radial basis function kernel is the default and works well in most situations [46].

SVM is well recognized in the field of machine learning and pattern recognition [47] and has recently been introduced in context of remote sensing image analysis [48]. SVM has been applied successfully for classifying multispectral imagery for example in [49] and outperformed other methods in the very most recent cases. In some studies SVM has been seen to produce higher accuracies than other classifiers, as a maximum likelihood classifier, an ANN and a simple decision tree [48]. Different voting schemes for multi class SVMs have been explored [48].

Post-classification refinements were enforced to diminish categorization errors as a result of the similarities in spectral responses of certain training classes such as bare fields and urban areas and some crop fields and wetlands. Independent samples of about 100 pixels for each class were randomly selected from each classifica tion category to assess classification accuracies. Confusion matrices as cross-tabulations of the mapped classversus the reference class were used to assess classification accuracies [50]. Overall accuracy, user's and producer's accuracies, and the Kappa statistic were then derived from the error matrices. The Kappa statistic incorporates the off diagonal elements of the confusion matrices (i.e., classification errors) and represents agreement obtained after removing the proportion of agreement that could be expected to occur by chance. Following the classification of imagery from the individual years, a GIS based multi-date post classification coeparison (PCC) change detection strategy was employed to determine changes in land-use in Nakuru. Change detection analysis entailed finding the type, amount and location of land use changes that had taken place.

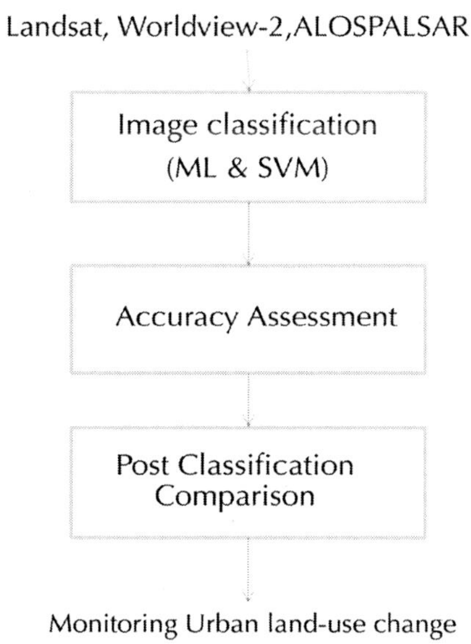

Figure 5: Image workflow for classification and change detection of multi-sensor data sets.

RESULTS AND DISCUSSION

Image classification of Landsat for 1986, 2000 and 2010 with 28.5 meters (original) spatial resolution was successfully performed and the results tabulated in Table 1. SVM produced better results compared to ML for example SVM in 2010 yielded an overall accuracy of 86.96% compared to ML of the same year with 83.80%. Land-use maps for Nakuru municipality for the three years are illustrated on Figure 6.

Furthermore the Radial Basis Function (RBF) kernel of support vector machine at various percentages was explored. RBF performed best at 0.33 compared to 0.5 and 0.7. The RBF value of 0.33 is the default value in ENVI software. For example the overall accuracy value of 93.42% was obtained for the year 1986 at RBF value of 0.33 compared to 93.14% at 0.5 and 93.14% at 0.7. Similarly an overall accuracy value of 83.77% was obtained for the year 2000 at RBF value of 0.33 compared to 82.75% at 0.5 and 82.75% at 0.7.

Land-use summary for Nakuru municipality (using Landsat at spatial resolution of 28.5 meters) was performed in ENVI and results tabulated in Table 2. SVM gave better results compared to ML. For example, looking at the urban class, using SVM growth was from 4.78 km^2 in 1986, 12.73 km^2 in 2000 and 27.84 km^2 in 2010 compared to ML.

Next we analyzed the upscaling of World view-2 and ALOS for Nakuru CBD which yielded low classification accuracy as shown in Table 3. World view-2 was up scaled from 0.5 to 28.5 meters and ALOS PALSAR from 12.5 to 28.5 meters.

Land-use summary for Nakuru CBD, using all data sets at spatial resolution of 28.5 meters, was performed in ENVI and results tabulated in Table 4. From Table 4 the respective SVM values of Landsat 2010, Worldview-2 and ALOS PALSAR for the three land-uses were different because of the different sizes of the training polygons and heterogeneity of the individual land-use classes.

Land-use summary for Nakuru CBD, using all data sets at spatial resolution of 12.5 meters, was performed in ENVI and results tabulated in Table 5. Relatively good values of urban class were achieved using SVM ALOS data as compared to Landsat data and World view-2 at 12.5 meters spatial resolution. Land-use maps for Nakuru CBD are illustrated in Figures 7-10.

Table 1: Confusion matrix for land-cover classification for nakuru municipality (landsat data from 1986, 2000 & 2010)

Method		Landsat
	Overall	
SVM	Accuracy	93.42%83.77% 86.96%
	Kappa	
	Coefficient	0.90870.79880.8393
	Overall	
	Accuracy	89.00%82.50% 83.80°O
ML	Kappa	
	Coefficient	0.85270.78620.8025

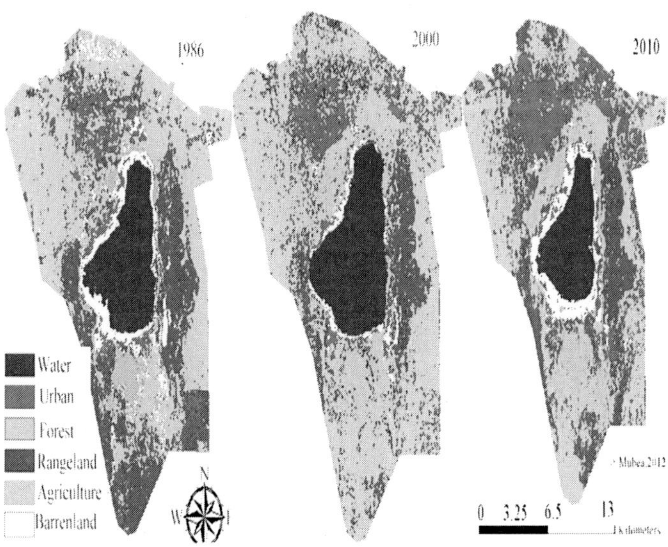

Figure 6: Land-use map for Nakuru municipality from Landsat for 1986, 2000 and 2010.

From the Table 6, downscaling of Landsat data sets did not affect classification accuracy. Landsat was downscaled from 28.5 meters to 12.5 meters. For example, the kappa coefficient value of 96.18% in Table 3 for Landsat 1986 classification is close to that one of 96.23% in Table 6 of the same data set classification using SVM.

The ratio of HH/HV ALOS PALSAR was combined with (bands 2, 3, 4) Landsat 2010 as a layer stack and SVM carried out. The combined ALOS PALSAR (2010) and Landsat (2010) gave better results in terms of classification accuracies. For example an overall accuracy value of 93.64 % and kappa statistic 0.8795 was obtained using the combined Landsat and ALOS PALSAR compared to that of ALOS PALSAR alone 84.09% and 0.6878 respectively as illustrated on Table 6 using SVM.

Based on the results from the different image classification, land-use change was calculated using post classification technique and results tabled in Table 7 and maps illustrated on Figures 11 and 12. From Table 7, 6.25 km^2 of land changed to urban land-use from non-urban between the years 1986 to 2000 compared to the value of 19.70 km^2 between the years 2000 to 2010. Non-urban land-use included forest and agriculture land-uses. Thus rapid urban growth has taken place between the years 2000 to 2010 due to urbanization. This has been witnessed by the high population growth [27,28], and rapid infrastructure developments such as more houses to cater for the growing demand for housing

CONCLUSIONS

Monitoring urban land-use and spatial-temporal changes is essential for guiding decision making in resource management. We analyzed land-use changes between 1986 and 2010 using multi-sensor satellite datasets. The performances of various classifiers were assessed. SVM algorithm gave better results compared to ML classifier. Image upscaling and downscaling were explored in this research. Downscaling of Landsat data sets did not affect classification accuracy as compared to upscaling of Worldview-2 and ALOS PALSAR datasets. The integration of Landsat and ALOS PALSAR yielded good results compared to when ALOS PALSAR was classified alone. Post classification comparison was performed to determine the land-use changes and conversion.

Table 2: Land-use estimates for Nakuru municipality using Landsat (28.5 spatial resolution)

	1986		2000		2010	
Land over classes (km2)	MLC	SVM	MLC	SVM	MLC	SVM
Water	33.29	36.95	38.77	40.09	27.25	30.78
Urban	16.65	4.78	26.4	12.73	26.83	27.84
Forest	16.95	20.55	25.77	27.1	21.91	26.93
Barren land	24.69	13.46	9.5	5.98	18.82	19.84
Rangeland	93.63	150.07	70.44	134	111.15	122.57
Agriculture	104.78	64.2	119.12	70.1	84.03	62.03
Total	290	290	290	290	290	290

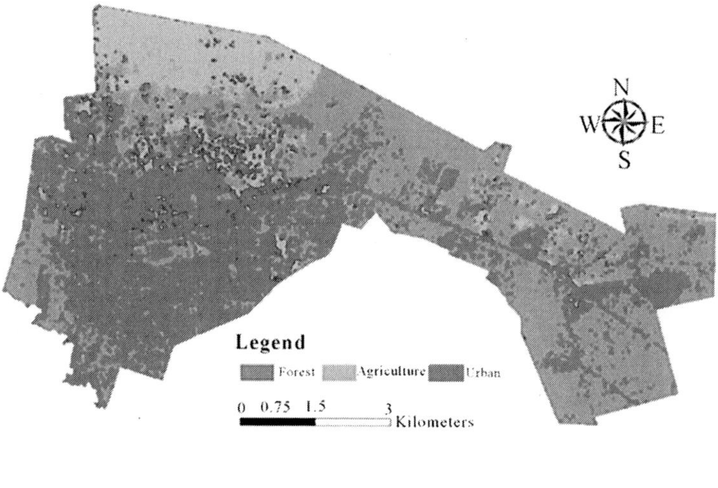

Figure 7: Land-use map for Nakuru CBD using Landsat TM 1986.

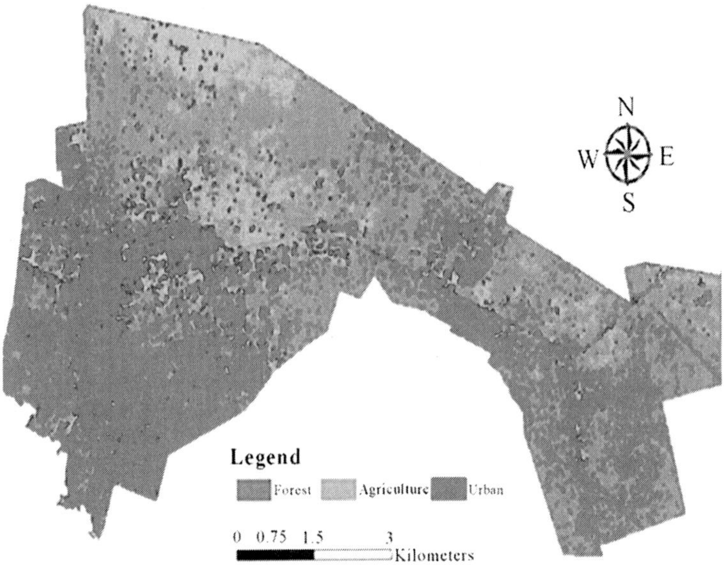

Figure 8: Land-use map for Nakuru CBD using Landsat ETM 2000.

Figure 9: Land-use map for Nakuru CBD using Landsat ETM+ 2000.

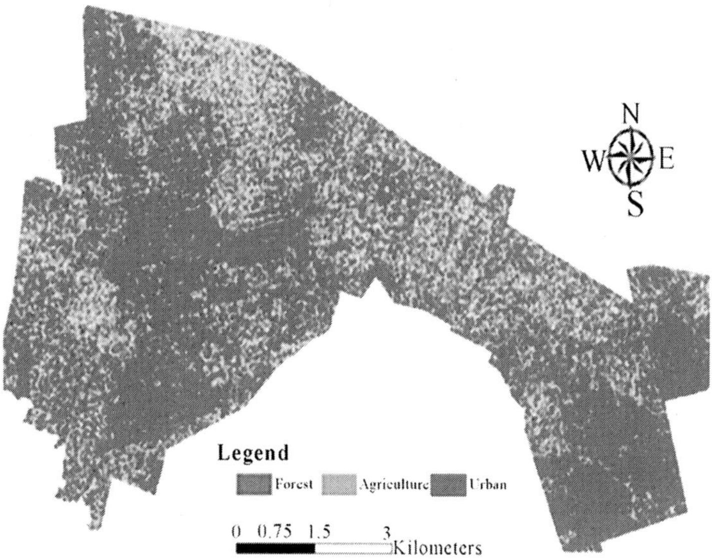

Figure 10: Land-use map for Nakuru CBD using ALOS PALSAR 2010.

Table 3: Comparison of classification accuracies for the three multi-sensor data sets at 28.5 spatial resolution for Nakuru CBD

Method			Landsat		Worldview -2	Alos palsar
		1986	2000	2010	2010	2010
	Overall Accuracy	96.18%	93.22%	99.51%	76.76%	59.37%
	Kappa Coefficient	0.9343	0.7321	0.9914	0.6358	0.3555
SVM						
	Overall Accuracy	95.24%	93.22%	99.60%	76.76%	18.62%
ML						
	Kappa Coefficient	0.9196	0.7321	0.9929	0.6358	0.0000

Table 4: Land-use estimates for Nakuru CBD (28.5 spatial resolution)

Land-cover classes (km2)		Landsat		Worldview -2	ALOS PALSAR
	1986	2000	2010	2010	2010
	SVM	SVM	SVM	SVM	SVM
Forest	11.28	6.45	12.15	16.1	11.26
Urban	34.35	30.96	30	22.25	17.38
Agriculture	12.37	20.59	15.85	19.64	29.36
Total	58	58	58	58	58

Table 5: Land-use estimates for Nakuru CBD (12.5m spatial resolution)

Land-cover	Landsat			Worldview -2	ALOS PALSAR
classes (km2)	1986	2000	2010	2010	2010
	SVM	SVM	SVM	SVM	SVM
Urban	15.91	18.89	23.31	23.55	27.76
Forest	8.94	9.23	9.67	14.1	7.07
Agriculture	33.15	29.88	25.02	20.35	23.17
Total	58	58	58	58	58

The information obtained however, is very useful for planning purposes and for the appropriate allocation of resources and demonstrate the potential of combining multi-sensor Landsat data, Worldview-2, and PALSAR data to provide precise means to analyze and map changes in urban land-use. Moreover, such information can be used as inputs to decision support systems, policy decisions and land management.

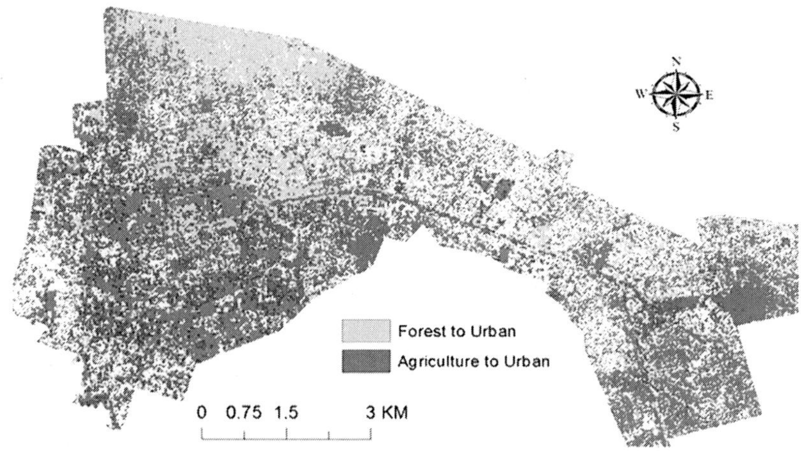

Figure 11: Land-use change in Nakuru CBD using Landsat TM (1986) and Landsat ETM (2000).

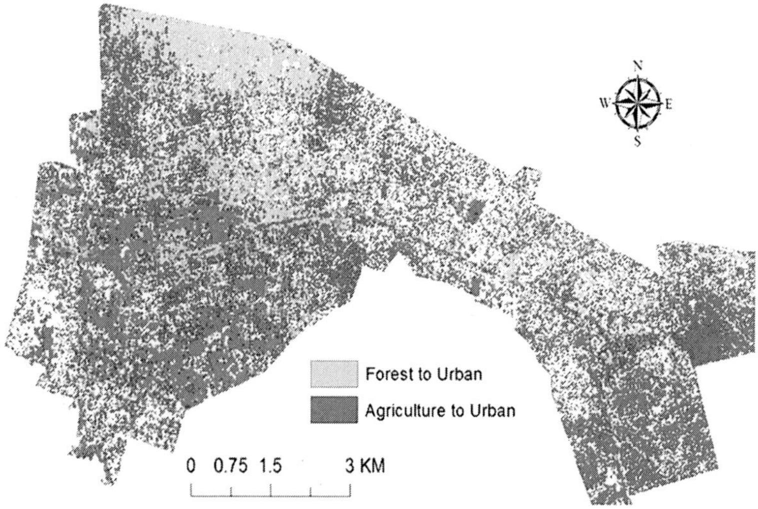

Figure 12: Land-use change in Nakuru CBD using Landsat ETM (2000) and ALOS PALSAR (2010).

Table 6: Confusion matrix for land-use classification for Nakuru CBD (12.5 spatial resolution)

			Landsat		Worldview -2	ALOS PALSAR
		1986	2000	2010	2010	2010
	Overall Accuracy	96.23%	96.91%	99.05%	77.47%	84.09%
SVM						
	Kappa Coefficient	0.943	0.9507	0.9836	0.6343	0.6878
	Overall Accuracy	96.66%	97.09%	98.68%	77.47%	21.00%
ML						
	Kappa Coefficient	0.9495	0.9485	0.9774	0.6343	0.0000

Table 7: Land-use change estimates based on post classification comparison

	1986 to 2000	2000 to 2010
Land-use change	Change (Km2)	Change (Km2)
No change	36.69	23.69
From non-urban to urban	6.25	19.70
from urban to non-urban	8.93	7.75

ACKNOWLEDGEMENTS

The authors would like to acknowledge European Space Agency (ESA) for providing ALOS PALSAR data and the Regional Centre for Mapping of Resources for Development (RCMRD, in Nairobi, Kenya) for providing Landsat 2010 and Worldview-2 which was significant for this research. Additionally the authors are grateful to German Academic Exchange Service (DAAD) and National Council for Science and Technology (NCST in Nairobi, Kenya) for providing the stipend to facilitate this research. Furthermore the authors acknowledge the contribution of the anonymous reviewers for their useful comments.

REFERENCES

1. UNECE, "Trends in Europe and North America," The Statistical Yearbook of the Economic Commission for Europe, 2003.
2. C. N. Mundia and M. Aniya, "Modeling Urban Growth of Nairobi City Using Cellular Automata and Geographical Information Systems," Geographical Review of Japan, Vol. 80, No. 12, 2007, pp. 777-788. doi:10.4157/grj.80.777
3. C. Lavalle, L. Demichili, M. Turchini, P. Casals Carrsco and M. Niederhuber, "Monitoring Mega-Cities: The MURBANDY/ MOLAND Approach," Development in Practice, Vol. 11, No. 2-3, 2001, pp. 350-357. doi:10.1080/09614520120056478
4. J. I. Barredo and L. Demicheli, "Urban Development in Mega Cities: Modeling and Predicting Future Urban Growth," Cities, Vol. 20, No. 5, 2003, pp. 297-310.doi:10.1016/S0264-2751(03)00047-7
5. UN-HABITAT, "State of the World Cities 2010/2011," 2010. http://www.unhabitat.org/documents/SOWC10/R4.pdf
6. C. N. Mundia, M. Aniya and Y. Murayama, "Modeling Spatial Processes of Urban Growth in an African City. A Case Study of Nairobi," Nova Science Publishers, New York, 2010.
7. K. C. Clarke, B. O. Parks and M. P. Crane, "Geographic Information Systems and Environmental Modeling," Prentice Hall, Upper Saddle River, 2002.
8. C. N. Mundia and M. Aniya, "Dynamics of Land-Use/ Cover Changes and Degradation of Nairobi City, Kenya," Land Degradation and Development, Vol. 17, No. 1, 2006, pp. 97-108. doi:10.1002/ldr.702
9. G. Tóth, "Impact of Land-Take on the Land Resource Base for Crop Production in the European Union," Science of the Total Environment, Vol. 435-436, 2012, pp. 202-214.doi:10.1016/j.scitotenv.2012.06.103
10. B. Ghimire, J. Rogan and J. Miller, "Contextural LandCover Classification: Incorporating Spatial Dependence in Land-Cover

Classification Models Using Random Forests and the Getis Statistic," Remote Sensing Letters, Vol. 1, No. 1, 2010, pp. 45-54. doi:10.1080/01431160903252327

11. M. Pesaresi, "Texture Analysis for Urban Pattern Recognition Using Fine Resolution Panchromaticsatellite Imagery," Geographical and Environmental Modelling, Vol. 4, No. 1, 2000, pp. 43-63. doi:10.1080/136159300111360

12. F. Dell'Acqua and P. Gamba, "Discriminating Urban Environments Using Multiscale Texture and Multiple SAR Images," International Journal of Remote Sensing, Vol. 27, No. 18, 2006, pp. 3797-3812. doi:10.1080/01431160600557572

13. A. Puissant, J. Hirsch and C. Weber, "The Utility of Texture Analysis to Improve Per-Pixel Classifications for High to Very High Spatial Resolution Imagery," International Journal of Remote Sensing, Vol. 26, No. 4, 2005, pp. 733-745.doi:10.1080/01431160512331316838

14. J. Rogan and D. M. Chen, "Remote Sensing Technology for Mapping and Monitoring Land-Cover and Land-Use Change," Progress in Planning, Vol. 61, No. 4, 2004, pp. 301-325. doi:10.1016/S0305-9006(03)00066-7

15. C. T. Chen, K. S. Chen and J. S. Lee, "The Use of Fully Polarimetric Information for the Fuzzy Neural Classification of SAR Image," Transactions on Geoscience and Remote Sensing, Vol. 41, No. 9, 2003, pp. 352-365.

16. R. J. Dekker, "Texture Analysis and Classification of ERS SAR Images for Map Updating of Urban Areas in the Netherland," Geoscience and Remote Sensing, Vol. 41, No. 9, 2003, pp. 1950-1958. doi:10.1109/TGRS.2003.814628

17. T. M. Pellizzeri, P. Gamba, P. Lombardo and F. D. Acqua, "Multitemporal/Multi-Band SAR Classification of Urban Areas Using Spatial Analysis: Statistical versus Neural Kernel-Based Approach," Geoscience and Remote Sensing, Vol. 41, No. 10, 2003, pp. 2338-2353.doi:10.1109/TGRS.2003.818762

18. Z. Zhu, C. E. Woodcock, J. Rogan and J. Kellndorfer, "Assessment of Spectral, Polarimetric, Temporal, and Spatial Dimensions for Urban and Peri-Urban Land Cover Classification Using Landsat and SAR Data," Remote Sensing of Environment, Vol. 117, 2011, pp. 72-82. doi:10.1016/j.rse.2011.07.020

19. D. Amarsaikhan and T. Douglas, "Data Fusion and Multisource Image Classification," International Journal of Remote Sensing, Vol. 25, No. 14, 2004, pp. 3529-3539. doi:10.1080/0143116031000115111

20. C. Corbane, J. Faure, N. Baghdadi, N. Villeneuve and M. Petit, "Rapid Urban Mapping Using SAR/Optical Imagery Synergy," Sensors, Vol. 8, No. 11, 2008, pp. 7125- 7143.doi:10.3390/s8117125

21. B. Waske and S. van der Linden, "Classifying Multilevel Imagery from SAR and Optical Sensors by Decision Fusion," IEEE Transaction on Geoscience and Remote Sensing, Vol. 46, No. 5, 2008, pp. 1457-1466. doi:10.1109/TGRS.2008.916089

22. M. J. Barnsley and S. L. Barr, "A Graph Based Structural Pattern Recognition System to Infer Urban Land-Use from Fine Spatial Resolution Land-Cover Data," Computer, Environment and Urban Systems, Vol. 21, No. 3-4, 1997, pp. 209-225.doi:10.1016/S0198-9715(97)10001-1

23. J. Mas, A. Pérez-Vega and K. C. Clarke, "Assessing Simulated Land-Use/Cover Maps Using Similarity and Fragmentation Indices," Ecological Complexity, Vol. 11, 2012, pp. 38-45.doi:10.1016/j.ecocom.2012.01.004

24. Republic of Kenya, "Kenya Population Census 1969," Nairobi, 1970.

25. Republic of Kenya, "Kenya Population Census 1979," Nairobi, 1981.

26. Republic of Kenya, "Kenya Population Census 1989," Nairobi, 1994.

27. Republic of Kenya, "Economic Survey 2000," Nairobi, 2000.

28. W. Mwangi, "Kenya Safaris-Nakuru Travel," University of Amsterdam, Amsterdam, 2007.

29. C. O. Obura, "Towards an Environmental Planning Approach in Urban Industrial Sitting and Operations in Kenya: The Case of Eldoret Town," Netherlands Geographical Studies, Utrecht, 1996.

30. M. Herold, D. A. Roberts, M. E. Gardner and P. E. Dennison, "Spectrometry for Urban Area Remote Sensing Development and Analysis of a Spectral Library from 350 to 2400 nm," Remote Sensing of Environment, Vol. 91, No. 3-4, 2004, pp. 304-319. doi:10.1016/j.rse.2004.02.013

31. C. Small, "A global Analysis of Urban Reflectance," International Journal of Remote Sensing, Vol. 26, No. 4, 2005, pp. 661-681. doi:10.1080/01431160310001654950

32. P. Griffiths, P. Hostert, O. Gruebner and S. Linden, "Mapping Megacity Growth with Multi-Sensor Data," Remote Sensing of Environment, Vol. 114, No. 2, 2010, pp. 426- 439.doi:10.1016/j.rse.2009.09.012

33. S. Phinn, M. Stanford, P. Scarth, A. T. Murray and P. T. Shyy, "Monitoring the Composition of Urban Environments Based on the Vegetation-Impervious Surface-Soil (VIS) Model by Subpixel Analysis Techniques," International Journal of Remote Sensing, Vol. 23, No. 22, 2002, pp. 4131-4153. doi:10.1080/01431160110114998

34. F. Pacifici, F. Del Frate, W. J. Emery, P. Gamba and J. Chanussot, "Urban Mapping Using Coarsesar and Optical Data," IEEE Geoscience and Remote Sensing Letters, Vol. 5, No. 3, 2008, pp. 331-335. doi:10.1109/LGRS.2008.915939

35. L. Gomez-Chova, et al., "Urban Monitoring Using MultiTemporal SAR and Multi-Spectral Data," Pattern Recognition Letters, Vol. 27, No. 4, 2006, pp. 234-243.doi:10.1016/j.patrec.2005.08.004

36. NASA, "Landsat Data Continuing Mission," 2008. http://ldcm.nasa.gov/index.html

37. M. A. Wulder, et al., "Landsat Continuity: Issues and Opportunities for Land Cover Monitoring," Remote Sensing of Environment, Vol. 112, No. 3, 2008, pp. 955-969.doi:10.1016/j.rse.2007.07.004

38. F. Garestier, P. Dubois-Fernandez, X. Dupuis, P. Paillou and I. Hajnsek, "PolInSAR Analysis of X-Band Data over Vegetated and Urban Areas," IEEE Transactions on Geoscience and Remote Sensing, Vol. 44, No. 2, 2006, pp. 356-364. doi:10.1109/TGRS.2005.862525

39. F. M. Henderson and A. J. Lewis, "Principles and Applications of Imaging Radar," 3rd Edition, John Wiley & Sons, New York, 1998.

40. Y. Dong, B. Forster and C. Ticehurst, "Radar Backscatter Analysis for Urban Environments," International Journal of Remote Sensing, Vol. 18, No. 6, 1997, pp. 1351-1364. doi:10.1080/014311697218467

41. R. Z. Schneider, K. P. Papathanassiou, I. Hajnsek and A. Moreira, "Polarimetric and Interferometric Characterization of Coherent

Scatterers in Urban Areas," IEEE Transactions on Geoscience and Remote Sensing, Vol. 44, No. 4, 2006, pp. 971-983.doi:10.1109/TGRS.2005.860950

42. J. F. Anderson, E. E. Hardy, J. T. Roach and R. E. Witmer, "A Land-Use and Land Cover Classification System for Use with Remote Sensor Data," U.S. Geological Survey, Washington, 1976.
43. N. Lam and D. A. Quattrochi, "On the Issues of Scale, Resolution, and Fractal Analysis in the Mapping Sciences," Professional Geographer, Vol. 44, 1992, pp. 88-98.doi:10.1111/j.0033-0124.1992.00088.x
44. D. P. Ming, J. Y. Yang, L. X. Li and Z. Q. Song, "Modified ALV for Selecting the Optimal Spatial Resolutionand Its Scale Effect on Image Classification Accuracy," Mathematical and Computer Modelling, Vol. 54, No. 3-4, 2011, pp. 1061-1068.doi:10.1016/j.mcm.2010.11.036
45. T. Idol, B. Haack, S. Sawaya, and A. Sheoran, "Land Cover/Use Mapping with Quad Polarization Radar and Derived Texture Measures," American Society of Photogrammetry and Remote Sensing, Oregon, 2008.
46. H. Z. M. Shafri and F. S. H. Ramle, "A Comparison of Support Vector Machine and Decision Tree Classifications Using Satellite Data of Langkawi Island," Journal of Information Technology, Vol. 8, No. 1, 2009, pp. 64-70. doi:10.3923/itj.2009.64.70
47. B. Waske and J. A. Benediktsson, "Fusion of Support Vector Machines for Classification of Multisensor Data," IEEE Transaction on Geoscience and Remote Sensing, Vol. 45, No. 12, 2007, pp. 3858-3866. doi:10.1109/TGRS.2007.898446
48. F. Melgani and L. Bruzzone, "Classification of Hyperspectral Remote Sensing Images with Support Vector Machines," IEEE Transactions on Geoscience and Remote Sensing, Vol. 42, No. 8, 2004, pp. 1778-1790. doi:10.1109/TGRS.2004.831865
49. C. Huang, L. S. Davis and J. R. G. Townshend, "An Assessment of Support Vector Machines for Land Cover Classification," International Journal of Remote Sensing, Vol. 23, No. 4, 2002, pp. 725-749. doi:10.1080/01431160110040323
50. R. G. Congalton and K. Green, "Assessing the Accuracy of Remotely Sensed Data: Principles and Practices," Lewis Publishers, Boca Raton, 1999.

Citations

CHAPTER 1

Niles Lehman, Tess Bernhard, Brian C Larson, Andrew JN Robinson, and Christopher CB Southgate, Empirical Demonstration of Environmental Sensing in Catalytic RNA: Evolution of Interpretive Behavior at the Origins of Life, doi:10.1186/s12862-014-0248-2.

CHAPTER 2

Sarah E Giuliani, Ashley M Frank, Danielle M Corgliano, Catherine Seifert, Loren Hauser, and Frank R Collart, Environment Sensing and Response Mediated by ABC Transporters, doi:10.1186/1471-2164-12-S1-S8.

CHAPTER 3

Ahmed S Khan and Sanie SS Sesay, Seafood Insecurity, Bush Meat Consumption, and Public Health Emergency in West Africa: Did We Miss the Early Warning Signs of an Ebola Epidemic?, doi:10.1186/s40152-015-0020-2.

CHAPTER 4

Omar Said and Alaa Elnashar, Scaling of Wireless Sensor Network Intrusion Detection Probability: 3D Sensors, 3D Intruders, and 3D Environments, doi: 10.1186/s13638-015-0240-6.

CHAPTER 5

Satoshi Nakada, Katsuhisa Baba, Masatoshi Sato, Masafumi Natsuike, Yoichi Ishikawa, Toshiyuki Awaji, Koji Koyamada, and Sei-Ichi Saitoh, The Role of Snowmelt Runoff on the Ocean Environment and Scallop Production in Funka Bay, Japan, doi:10.1186/s40645-014-0025-2.

CHAPTER 6

Ines Mack, Andreas Hector, Marlene Ballbach, Julius Kohlhäufl, Katharina J Fuchs, Alexander Weber, Marcus A Mall, and Dominik Hartl, The Role of Chitin, Chitinases, and Chitinase-like Proteins in Pediatric Lung Diseases, doi:10.1186/s40348-015-0014-6.

CHAPTER 7

Jaqueline Daniel-Gromke, Jan Liebetrau, Velina Denysenko, and Christian Krebs, Digestion of Bio-waste - GAG Emissions and Mitigation Potential, doi: 10.1186/s13705-014-0032-6.

CHAPTER 8

K. Mubea and G. Menz, "Monitoring Land-Use Change in Nakuru (Kenya) Using Multi-Sensor Satellite Data,"Advances in Remote Sensing, Vol. 1 No. 3, 2012, pp. 74-84, doi: 10.4236/ars.2012.13008.

Index